Biochemical and Nutritional Changes during Food Processing and Storage

Biochemical and Nutritional Changes during Food Processing and Storage

Editors

Vibeke Orlien
Tomas Bolumar

MDPI • Basel • Beijing • Wuhan • Barcelona • Belgrade • Manchester • Tokyo • Cluj • Tianjin

Editors
Vibeke Orlien
Department of Food Science,
University of Copenhagen
Denmark

Tomas Bolumar
Max Rubner Institute,
German Federal Research
Institute of Nutrition and Food,
Department of Safety and
Quality of Meat
Germany

Editorial Office
MDPI
St. Alban-Anlage 66
4052 Basel, Switzerland

This is a reprint of articles from the Special Issue published online in the open access journal *Foods* (ISSN 2304-8158) (available at: https://www.mdpi.com/journal/foods/special_issues/storage_process_nutritional).

For citation purposes, cite each article independently as indicated on the article page online and as indicated below:

LastName, A.A.; LastName, B.B.; LastName, C.C. Article Title. *Journal Name* **Year**, *Article Number*, Page Range.

ISBN 978-3-03943-416-9 (Hbk)
ISBN 978-3-03943-417-6 (PDF)

Contents

About the Editors

Vibeke Orlien (Associate Professor) holds an MSc in Chemistry and Mathematics (1997) from Roskilde University and a PhD in Food Chemistry (2003) from the Department of Food Science of the Royal Veterinary and Agricultural University. In 2007, the Royal Veterinary and Agricultural University merged with the University of Copenhagen. Prof. Orlien was Head of the Food Chemistry section in the Department of Food Science (2010–2016). In 2016, the department reorganized, and she is currently Deputy Head of Section and Deputy Manager of the Food Design and Consumer Behavior section. Prof. Orlien has a wide range of experience in management and coordination of research projects, both basic science and application-oriented projects with industry, including supervision of postdocs and PhD students. Her research areas are thermodynamics in food chemistry, non-thermal processing technologies, and protein modification. She has particular research interest in process–property–structure interactions in food materials and food systems from molecular size to macroscopic levels based on molecular and mechanistic understanding. Her work also focuses on practical functional food design, covering the coupling between food process methods and technologies and food chemistry. She has more than 60 peer-reviewed publications (h-index: 24 Web of Science/26 Scopus).

Tomas Bolumar (Research Scientist) holds a BSc in Food Science and Technology (1998) from the University of Valencia and a PhD in Meat Biochemistry (2004) from the Institute of Agrochemistry and Food Technology of the Spanish High Scientific Research Council (IATA-CSIC). He is currently Head of the Meat Technology section at the Max Rubner Institute (Germany). He has a background in biochemistry and processing of fresh and further-processed meats, with more than 20 years' experience in the transformative unit operations along the entire post-harvest value chain of meat (from abattoir to consumer). After his PhD, Dr. Bolumar worked in industrial R&D. Later on, he gained extensive international experience by working in basic and applied science with different companies at leading research centers such as the University of Copenhagen (Denmark), the German Institute of Food Technology (DIL, Germany) and the Commonwealth for Scientific and Industrial Research Organization (CSIRO, Australia). In the last decade, he has carried out research in collaborative interdisciplinary scenarios at the interfaces of meat quality and advanced food processing methods, pioneering the progress of disruptive technologies in meat processing such as shockwave, high pressure processing, pulsed electric fields and cutting-edge automation systems. Some of his research outcomes have been condensed in review articles among his > 50 publications (h-index=19). His R&D activities cover the impact of emerging processing methods on intrinsic biochemical and physical properties of meat products, also considering the relationship of these processing methods with prime safety and quality attributes. His goal is to optimize meat processing to better fit the needs of industry and society and further the development of meat products with enhanced nutritional profiles.

Editorial

Biochemical and Nutritional Changes during Food Processing and Storage

Vibeke Orlien [1,*] **and Tomas Bolumar** [2]

[1] Department of Food Science, University of Copenhagen, Rolighedsvej 26, DK-1958 Frederiksberg, Denmark
[2] Department of Safety and Quality of Meat, Max Rubner Institute, Federal Research Institute of Nutrition and Food; E.-C.-Baumann- Straße 20, 95326 Kulmbach, Germany; tomas.bolumar@mri.bund.de
* Correspondence: vor@food.ku.dk; Tel.: +45-353-332-26

Received: 11 October 2019; Accepted: 11 October 2019; Published: 14 October 2019

1. Introduction

Domestic food processing goes a long way back in time, for example, heat for cooking was used 1.9 million years ago. Nowadays, food and meal preparation seems to be moving out of the home kitchen into factories, and pre-processed or processed/convenience foods are becoming a larger part of the daily diet. In addition, consumers are progressively focusing on the impact of their food on health, and demand foods that have a high nutritional quality, and aroma and natural flavor similar to freshly-made products. Therefore, nutritional quality is concurrent with food safety, and sensory perception is becoming an increasingly important factor in food choices. The human digestive tract disintegrates food in order for nutrients to be released and be made available to the body. However, nutrients can undergo unwanted degradation upon processing and subsequent storage, negatively influencing the nutritional value of food and its physiological effects. Different processing techniques will result in different food structures, thereby also affecting bioaccessibility, bioavailability, and overall nutritional value. Hence, food scientists and industry have an increased interest in both conventional and innovative processing methods that can provide products with good quality and high nutritional value, along with a stable shelf life.

This Special Issue aims to shed some light on the latest knowledge and developments regarding the effects of food processing and storage on biochemical and nutritional changes.

2. Effect of Processing and Storage on Biochemical and Nutritional Changes

The processing of food raw material often targets specific compounds in order to achieve the desired texture and taste, for example, treatment of milk in order to coagulate proteins to form a cheese and subsequent storage to develop flavor. In complex matrices such as meat and vegetables, the processing will most likely also affect other compounds and hence produce biochemical changes that may affect the product properties in a negative manner. In meats, the formation of unwanted substances or safety concerns are sometimes related to processing conditions, for instance, the formation of polycyclic aromatic hydrocarbons during intensive particular smoking processes, the generation of heterocyclic aromatic amines in particular heating/ grilling conditions, and the release of compounds from the oxidation reactions of lipids and proteins that can affect flavour and texture [1,2]. The opposite is also possible, and one can favor processes that boost the presence of nutritional valuable components in meat products and/or facilitate digestion. For instance, through an extensive proteolysis, the formation of bioactive peptides with different bioactivity such as antioxidant, antihypertensive, immunomodulating, antimicrobial, prebiotic, and hypocholesterolemic properties can be enhanced in a variety of fermented and aged meat products [3]. The application of mild preservation processes can better preserve highly regarded nutrients such as vitamins.

Obviously, the product goal can be reached without extensive detrimental effects by selecting proper processing parameters. However, this is not a straightforward task, and more knowledge at the molecular level is still needed. This Special Issue covers a total of six articles (five research papers and one commentary report) concerning the effect of new processing technologies, packaging methods, protein extraction, and color ingredients on products or compounds related to biochemical and nutritional changes.

Juices are a well-known healthy food product, produced and enjoyed for centuries, but the juicing process is in fact allowing contact between degradative enzymes and healthy phenolic compounds leading to flavor and nutritional changes. The paper 'Influence of Pulsed Electric Field and Ohmic Heating Pretreatments on Enzyme and Antioxidant Activity of Fruit and Vegetable Juices' provides new and valuable information about the optimization of the pulsed electric field (PEF) and ohmic heating (OH) treatments for reducing the energy requirements and process time and increasing yield and quality. Thus, color, antioxidant activity (DPPH and ABTS method), and enzyme (peroxidase and polyphenoloxidase) activity were investigated in carrot and apple juices subjected to various PEF and OH treatments. In conclusion, both PEF and OH were found to positively contribute to improved juice quality by enhanced ingredient release and retention. Fish is also a well-known healthy food, but also highly perishable, and upon storage, protein oxidation and degradation have severe negative effects on nutritional value. The paper 'Effect of Different Packaging Methods on Protein Oxidation and Degradation of Grouper (Epinephelus coioides) During Refrigerated Storage' shows how different packaging methods such as air packaging (AP), vacuum packaging (VP), and modified atmosphere packaging (MAP) affect protein oxidation and degradation of grouper fillets during refrigerated storage. By monitoring changes in total sulfhydryl and disulfide bonds, carbonyl content and hydrophobicity, ATPase activity, soluble peptides, myofibril fragmentation index, free amino acids, protein secondary structure, and total protein electrophoresis, the authors showed that the degree of grouper fillet protein oxidation was increased upon storage. Moreover, it was found that protein oxidation and degradation were highly correlated. In conclusion, the high-carbon-dioxide MAP packaging method played a positive role in the inhibition of myofibril degradation and oxidation for refrigerated grouper fillets. Likewise, it is now well-known, for instance, that the use of rich oxygen MAP in other muscle foods such as red meat accelerates oxidative processes that affect color, flavor, and texture. Recent research has found that by application of proper antioxidant strategies and reducing the oxygen concentration in the package, it is possible to mitigate this problem and obtain a product of superior quality at the point of sale [4,5]. The two papers published in this special Issue are examples of how the investigation of mechanism at a molecular level can contribute to an overall assessment of applied technology and packaging on product quality and nutritional value. Both papers are generic in its methodology, which can be transferred to other raw food materials.

Color is an important quality attribute of food products for consumer acceptability. In this issue, two papers provide information about the stability of two natural colorants, namely carotenoids and anthocyanins, which both also have health promoting properties. The paper 'Stabilization of Crystalline Carotenoids in Carrot Concentrate Powders: Effects of Drying Technology, Carrier Material, and Antioxidants' seeks to stabilize carrot carotenoid crystals by spray- and freeze-drying, addition of functional additives, and oxygen free storage. An analytical approach was applied in order to qualitatively assess the physical state and the pigment concentration during production and storage. In conclusion, the exclusion of oxygen clearly had the most profound effect on carotenoid stability during storage. On the other hand, in the paper 'The Influence of Chemical Structure and the Presence of Ascorbic Acid on Anthocyanins Stability and Spectral Properties in Purified Model Systems' the influence of anthocyanins' structure, pH, and ascorbic acid on the stability and spectral properties of anthocyanins during simulated shelf life was investigated by spectral and high performance liquid chromatography-mass spectrometry analyses. The systematic stability study showed a higher stability in acidic medium and enhanced stability with increasing size of conjugated sugar, and a rapid and high anthocyanin degradation when stored without cooling and with the addition of ascorbic acid, which both should be avoided to protect anthocyanins from degradation.

The growing population has put a focus on the question of whether there is enough protein to feed the increasing number of humans and animals. Therefore, the examination of protein raw materials is of both scientific and applied interest. The paper 'Variations in Amino Acid and Protein Profiles in White versus Brown Teff (Eragrostis Tef) Seeds, and Effect of Extraction Methods on Protein Yields' compares the nutritional qualities by amino acid composition among six teff seed types and three different protein extraction methods. Maybe not surprisingly, a clear genetic variability between white and brown teff seed types was established. Interestingly, brown teff had higher content of essential amino acid than the white type. Moreover, the extraction method gave different results concerning the type of protein extracted and can thus be used to tune the quality or functional differences among teff protein fractions or meals.

The understanding of the digestibility of different foods can also play a decisive role in providing recommendations for nutritional guidelines and how to make the best use of the available food resources. The review paper 'The Effect of Processing on Digestion of Legume Proteins' highlights that protein digestibility increases after processing using different processing methods. However, since both the type of legume and the applied methods differed, it cannot be concluded which specific method is best for each individual legume type. Therefore, further research is required at the legume type level to provide processing recommendations that maximize bioavailability.

Biochemical and nutritional changes during food processing and storage have important implications for both consumer protection and health as well as food quality. We forecast many more studies coming up to address the optimization of processing conditions, possibly with the incorporation of novel mild processing methods and packaging processes as a way for food authorities and industry to minimize the presence of unwanted compounds and maximize the quality and nutritional value of food.

We would like to acknowledge all the authors and reviewers for their excellent contribution, effort, and comments.

Author Contributions: V.O. and T.B. wrote the paper.

Funding: This research received no external funding.

Conflicts of Interest: The authors declare no conflict of interest.

References

1. Flores, M.; Mora, L.; Reig, M.; Toldrá, F. Risk assessment of chemical substances of safety concern generated in processed meats. *Food Sci. Hum. Wellness* **2019**. [CrossRef]
2. Gibis, M.; Kruwinnus, M.; Weiss, J. Impact of different pan-frying conditions on the formation of heterocyclic aromatic amines and sensory quality in fried bacon. *Food Chem.* **2015**, *168*, 383–389. [CrossRef] [PubMed]
3. Arihara, K. Strategies for designing novel functional meat products. *Meat Sci.* **2006**, *74*, 219–229. [CrossRef] [PubMed]
4. Lund, M.N.; Lametsch, R.; Hviid, M.S.; Jensen, O.N.; Skibsted, L.H. High-oxygen packaging atmosphere influences protein oxidation and tenderness of porcine longissimus dorsi during chill storage. *Meat Sci.* **2007**, *77*, 295–303. [CrossRef] [PubMed]
5. Holman, B.W.B.; Kerry, J.P.; Hopkins, D.L. Meat packaging solutions to current industry challenges: A review. *Meat Sci.* **2018**, *144*, 159–168. [CrossRef] [PubMed]

Article

Effect of Different Packaging Methods on Protein Oxidation and Degradation of Grouper (*Epinephelus coioides*) During Refrigerated Storage

Xicai Zhang [1,2,3,4,5], **Wenbo Huang** [1,2,3,4] and **Jing Xie** [1,2,3,4,*]

1   College of Food Science & Technology, Shanghai Ocean University, Shanghai 201306, China
2   Shanghai Engineering Research Center of Aquatic Product Processing and Preservation, Shanghai 201306, China
3   Shanghai Professional Technology Service Platform on Cold Chain Equipment Performance and Energy Saving Evaluation, Shanghai 201306, China
4   National Experimental Teaching Demonstration Center for Food Science and Engineering, Shanghai Ocean University, Shanghai 201306, China
5   Jingchu University of Technology, Jingmen 448000, China
*   Correspondence: jxie@shou.edu.cn; Tel.: +86-021-6190-0391

Received: 13 June 2019; Accepted: 6 August 2019; Published: 7 August 2019

Abstract: This study investigates the effect of different packaging methods—namely, air packaging (AP), vacuum packaging (VP), and modified atmosphere packaging (MAP)—on the protein oxidation and degradation of grouper (*Epinephelus coioides*) fillets during refrigerated storage. The carbonyl group, myofibril fragmentation index, free amino acids, FTIR of myofibrillar proteins, and total protein SDS-PAGE were determined. The results showed that the protein oxidation degree of the fillets gradually increased as the storage time increased. The FTIR results indicated that the secondary structure transformed from an α-helix to an irregular curl. SDS-PAGE confirmed the degradation of the myosin heavy chain, and that myosin gradually occurred during refrigerated storage. Meanwhile, protein oxidation and degradation were highly correlated. Protein degradation was accelerated by protein oxidation in myofibrils, which included the increase of protein surface hydrophobicity and changes of the secondary structure. In fact, the protein oxidation and degradation of the grouper fillets were effectively inhibited by MAP and VP during refrigerated storage, and MAP (30% N_2 and 70% CO_2) had the best results.

Keywords: grouper; refrigerated storage; packaging methods; protein oxidation; protein degradation

---

1. Introduction

Grouper (*Epinephelus coioides*), which belongs to the order Perciformes and the family Serranidae, is a warm-water, offshore demersal fish that is referred to as "marine chicken" because of its considerable similarity in taste to chicken meat. With the development of artificial breeding and breeding technology, grouper has become an important economic fish along the coast of China [1]. Furthermore, the living standards of residents have been significantly enhanced as the economy has developed. Due to the quickening pace of life of the younger generation, fresh fish fillets are becoming the main marketing form of fish products [2,3]. However, grouper can easily decompose, because of its abundant nutrition, high water content, and excellent protease activity. Protein is one of the most important nutrients in aquatic products. During cold storage, the changing forms of protein mainly include protein oxidation and degradation. Carbonyl compounds formed by protein oxidation can change the cell structure of the myofibrillar protein, affecting the hydrophobicity index of fish protein [4,5]. Soon after fish die, the protein initially breaks down into many intermediates with the hydrolysis of endogenous protease [6,7]. In the later stages, the enzymes produced by microbial reproduction lead to protein degradation, resulting in spoilage of the fish. Therefore, the change of biochemical characteristics of proteins in aquatic products is an important reason for the deterioration of such products' quality during cold storage [8]. Refrigerated fish fillets need to be packaged to ensure their freshness and to meet the needs of different customers. Presently, the main packaging methods of grouper are air packaging (AP), vacuum packaging (VP), and modified atmosphere packaging (MAP). AP and MAP have been considered to be effective ways of preserving food, due to the excellent isolation of oxygen and food [9,10]. MAP, due to the bacteriostatic effect of CO_2 inducing anaerobic conditions into the packaging environment, can effectively reduce apparent changes of aquatic products [11]. In our previous study, it was found that a high concentration of CO_2 in MAP (30% N_2 and 70% CO_2) could prolong the shelf life of cold-stored grouper. Currently, research on the effects of VP and MAP on the quality of aquatic products mainly focuses on the quality attributes, shelf-life assessment, changes of proteins in aquatic products during storage, and the denaturation of proteins [12–14]. Also, trichloroacetic acid (TCA)-soluble peptides, the myofibril fragmentation index (MFI), free amino acids (FAAs), SDS-PAGE, and FTIR of myofibrils are usually evaluated during storage.

However, the relationship between protein oxidation and degradation is controversial. On the one hand, the hydrophobicity caused by protein oxidation increases protease recognition and the subsequent oxidation of protein degradation [15]. On the other hand, the polymer caused by protein oxidation between protein crosslinking could affect the further degradation of proteins, and is a poor substrate of protease [16]—something which has not been studied in research on grouper. It is not clear if MAP or VP can isolate oxygen as an effective method of food preservation and slow down the progress of protein oxidation in grouper. Therefore, the purpose of this study is to investigate the relationship between protein oxidation and degradation, and to compare the preservation effects of protein oxidation and degradation among AP, MAP, and VP refrigerated storage on grouper fillets.

2. Materials and Methods

2.1. Sampling

Fresh grouper (weight: 500 ± 50 g; length: 35 ± 5 cm) were purchased from the wholesale aquatic market of Luchao Port (Shanghai, China) and immediately transported to the laboratory in crushed ice within 0.5 h.

Based on the cleaning process, the head, bone, and skin of the grouper were removed and drained under refrigerated temperatures. Afterwards, the dorsal part of the fish fillets were cut to 2 × 2 × 2 cm and packaged in polyethylene bags (low-density polyethylene; relative density of 0.917–0.924, 30 × 20 cm). The samples were respectively packaged in AP (meaning they were exposed to air), VP, and MAP (with 30% N_2 and 70% CO_2). All the fillets samples were refrigerated at 4.0 ± 0.5 °C. Relative indexes were evaluated regularly at 0, 3, 6, 9, 12, 15, and 18 days. The AP group was terminated at the 12th day due to the deterioration of the grouper fillets.

2.2. Extraction of Myofibrillar Protein

The extraction of myofibrillar protein was carried out according to the method of Ogawa et al. [17]. Briefly, 2 g fillets were weighed and washed by a fivefold volume of Tris-HCl buffer (pH 7.0, 10 mmol/L). Then, a fivefold volume of KCl-Tris buffer was added to the abovementioned solutions, followed by homogenization in an ice bath for 90 s (12,000 rpm), with a brief pause in the middle of the homogenization process to prevent overheating. The homogenate was centrifuged three times at 5000× g for 10 min. Subsequently, a fivefold volume of 10 mmol/L Tris-HCl buffer (0.6 mol/L NaCl, pH 7.0) was added and centrifuged repeatedly for 10 min at 5000× g. The supernatant was a myofibrillar protein extract, which was stored at −80 °C in a refrigerator for further use.

Protein concentration was determined by the method of Abbey et al. [18]. Standard curves were prepared by BSA (Bull Serum Albumin). Protein solution (0.05 mL) was added to 3 mL of

Bradford reagent, mixed, and kept still for 10 min. The OD (optical density) value was measured by a spectrophotometer at 595 nm. At the same time, the following protein concentrations were determined by the same method.

2.3. Carbonyl Content

The carbonyl content was determined by following the procedures mentioned by Oliver et al. [19]. The myofibrillar protein extract was adjusted to a concentration of 5 mg/mL with phosphate buffer solution (pH 7.0), and incubated in 1 mL 0.01 mol/L 2, 4-dinitrophenylhydrazine solution at 37 °C for 30 min. Then, 3 mL of 20% trichloroacetic acid was added and centrifuged at $8500\times g$ for 5 min. The supernatant was removed and the precipitate was washed six times with an ethyl acetate and ethanol mixture solution (1:1, *v/v*). Finally, the precipitate was dissolved in 5 mL of guanidine hydrochloride solution (6 mol/L) and incubated for 15 min under a 37 °C water bath, which was centrifuged for 10 min at $8500\times g$. Finally, the absorbance of the supernatant was measured at 370 nm. The carbonyl content was expressed as nmol carbonyl/mg protein.

2.4. Surface Hydrophobicity

Surface hydrophobicity was measured as described by Chelh et al. [20]. The abovementioned extracted myofibrillar protein was adjusted to 1 mg/mL with phosphate buffer solution (pH 7.0). Two hundred milliliters of bromophenol blue (1 mg/mL) were mixed with 1 mL of protein solution, followed by constant oscillation for 15 min to react sufficiently. As for the control group, the same procedure was implemented using phosphate solution to replace the extracted myofibrillar protein. Then, centrifugation was done at 4 °C and $2000\times g$ for 15 min. The supernatant was diluted 10 times, the absorbance of which was measured at 595 nm. Surface hydrophobicity was expressed by the following formula: bromophenol blue/mg protein = 200 μg × (OD control − OD sample)/OD control.

2.5. Total Sulfhydryl and Disulfide Bond Content

Total sulfhydryl and disulfide bond content was evaluated according to the method of Benjakul et al. [21]. One milliliter of myofibrillar protein solution (4 mg/mL) was added to 9 mL 0.2 mol/L Tris-HCl buffer (pH 6.8, 8 mol/L urea, 2% SDS, and 10 mmol/L EDTA (Ethylene Diamine Tetraacetic Acid)). Then, 0.4 mL 5,5-dithio-bis (2-nitrobenzoic acid) (0.1%) was added to the resulting mixture and incubated for 25 min at 40 °C. The absorbance was measured at 412 nm. At the same time, 0.6 mol/L KCl as a blank was also subjected to this step. The extinction coefficient of $13,600\ \text{mol}^{-1}\ \text{cm}^{-1}$ was used to calculate the total -SH group content. The content of the disulfide bond was calculated according to the method described by Thannhauser et al. [22].

2.6. Ca^{2+} ATPase Activity

The Ca^{2+} ATPase activity was determined by the method described by Benjakul et al. [23]. The content of inorganic phosphate in the supernatant was determined by the method of Thanonkaew et al. [24]. The activity of Ca^{2+} ATPase was defined as the milliliter (nmol/mg protein) of inorganic phosphate produced by 1 mg of protein in 1 min.

2.7. Trichloroacetic Acid-Soluble Peptide Content

According to the method of Sriket et al. [25], a 3 g sample was added to a 27 mL TCA solution (5%), which was homogenized by a high-speed tissue homogenizer for 1 min at 12,000 rpm and then placed in an ice bath for 1 h. Repeated centrifugation was carried out at 4 °C and $5000\times g$ for 5 min. The result was expressed as umol tyrosine/g muscle.

2.8. Myofibril Fragmentation Index

The method of Culler et al. was used to evaluate the MFI [26]. Briefly, the concentration of protein solution was adjusted to 0.5 mg/mL; then, the absorbance was measured at 540 nm, and the MFI value was equal to the OD value multiplied by 200.

2.9. Free Amino Acid Content

Samples of free amino acid extracts were prepared according to the method of Yu et al. [27]. Mobile phase 1 of an automatic amino acid analyzer (L-8800, Hitachi Co. Ltd., Tokyo, Japan) consisted of the buffer of sodium citrate and citric acid; the pH of the mixed buffers were 3.2, 3.3, 4.0, and 4.9, respectively. Mobile phase 2 was prepared by 4% ninhydrin (*v/v*). The test parameters were as follows: column (4.6×150 mm, 7 μm); column temperature (50 °C); channels 1 and 2 flow rates (0.4 mL/min and 0.35 mL/min, respectively).

2.10. FTIR Measure

The grouper fillets were powdered with KBr after being freeze-dried (MINFAST04, TIANLI Executive and Administration Management, Beijing, China); then, the mixed sample was pressed into flakes. An FTIR (Nicolet iS5, Thermo Scientific Inc, Waltham, MA, USA) spectrometer was used for the measurements. Infrared spectra were recorded with 32 scans in the 400–4000 cm^{-1} range with a resolution of 4 cm^{-1}; also, the operating environment was set at 25 °C. The recorded spectra were analyzed by Omnic professional software (Omnic professional, v 9.2, Thermo Nicole Inc., Waltham, MA, USA), and Gaussian fitting was used to analyze the second-derivative spectrum in the range of 1600–1700 cm^{-1} by PeakFit software (v 412, Systat Software Inc., San Jose, CA, USA).

2.11. SDS-PAGE

First, 3 g of minced samples and 30 mL of 50 g/L SDS solution were homogenized at 85 °C. Then, the mixture was subjected to heat preservation for 1 h after high-speed homogenization for 5 min (12,000 r/min), and the supernatant was taken after $5000\times g$ for 20 min [28]. SDS-PAGE was performed at a 4–20% gradient, and a real-band, three-color, high-range protein marker purchased from Sangon Biotech (Sangon Biotech Co., Ltd., Shanghai, China) was adopted. A sensitive protein fast staining kit was used for staining, and the decolorized gel was scanned by a gel image scanning system after electrophoresis (GelDoc XR, Bio-Rad Inc., Hercules, CA, USA). Background subtraction, band matching, and optical density calculation were analyzed by Quantity One software (Quantity One 4.0, Bio-Rad Inc., USA).

2.12. Statistical Analysis

Three replicates were used for all samples with three parallel tests. SPSS 8.0 (SPSS, Chicago, IL, USA) was used for statistical analysis. Univariate ANOVA was used to determine the statistical differences in different treatment groups. Duncan's multiple range test was used to determine the significant difference between the averages ($p < 0.05$). The data were expressed as the mean and standard deviation (SD), and Origin 8.5 (OriginLab, Northampton, MA, USA) was used for illustration.

3. Results and Discussion

3.1. Changes in Total Sulfhydryl and Disulfide Bonds

Figure 1 shows the changes of sulfhydryl and disulfide bonds in each group. As shown in Figure 1, the content of the total sulfhydryl groups in each treatment group presented a downward trend with the extension of storage time, which is similar to trends found in other studies on grouper (He et al., 2018). The sulfhydryl content of the AP group decreased to 49.05 nmol/mg of protein on the sixth day, which was significantly different compared with the VP and MAP groups. In contrast, the differences between the VP and MAP groups were not significant. The sulfydryl content in the VP group was slightly lower than that in the MAP group after nine days.

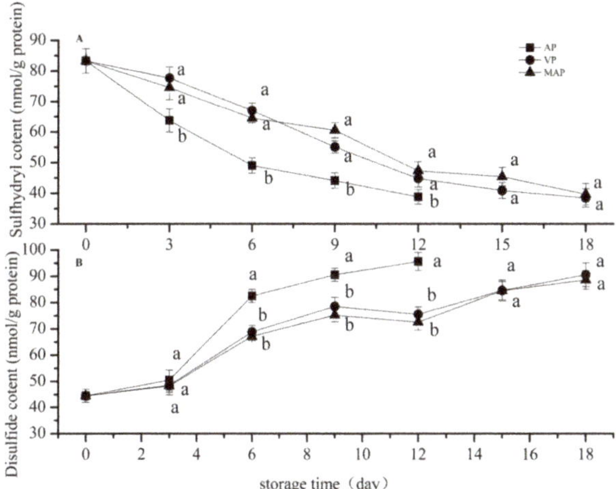

Figure 1. Changes in sulfhydryl (**top**) and disulfide (**bottom**) bond contents of grouper myofibrillar protein (AP: air packaging; VP: vacuum packaging; MAP: modified atmosphere packaging). Different lower-case letters in different groups from same day indicate a significant difference ($p < 0.05$).

The sulfhydryl group of myofibrils can be oxidized by reactive oxygen species (ROS) that produce disulfide bonds, sulfonic acid, and other oxidation products during refrigerated storage [29]. Protein oxidation can be reflected by the loss of the sulfhydryl group [30], which contains a disulfide bond; this is a common oxidation product that can gradually increase with the oxidation of the sulfhydryl group. The disulfide bond content in each group increased gradually with the oxidation of the sulfhydryl group (Figure 1), and the content of the AP group increased significantly, from 44.5 to 82.5 nmol/mg protein. It is worth noting that the disulfide bond contents of the AP and MAP groups decreased slightly starting from the ninth day, which may have been caused by the degeneration of myofibril.

As shown by the two-way analysis of variance, the storage time and packaging methods significantly affected the total sulfhydryl and disulfide bonds of the grouper fillets ($p < 0.01$). There was a significant difference between the VP and AP groups, and between the AP and MAP groups ($p < 0.01$); however, there was no significant difference between the VP and MAP groups ($p > 0.05$).

3.2. Changes in Carbonyl Content and Hydrophobicity

Protein carbonylation is one of the most important changes in the oxidation of muscle proteins, especially the extensive modifications caused by the oxidation of myofibrils. In particular, from the formation of carbonyl compounds [31], changes in the content of carbonyl compounds are usually used to represent the oxidation rate of proteins. In Figure 2, it can be seen that the change of carbonyl content in each group increased with the cold storage time. The initial carbonyl content of the fresh fish was increased 10 times in the AP group on day 6. The two-way analysis of variance revealed that protein carbonylation was significantly affected by storage time and packaging methods ($p < 0.05$).

The specific binding of bromophenol blue with myofibrillar is considered to be a simple and reliable method for the determination of surface hydrophobicity. The hydrophobicity of the protein surface can be determined by the degree to which bromophenol blue binds specifically to myofibrillar [20]. Due to the conformational changes induced by hydroxyl radicals, oxidized myofibrillar protein undergoes extensive exposure of hydrophobic groups [32]. This phenomenon was confirmed in our experiment. The degree to which bromophenol blue bound specifically to myofibrillar became serious in all three groups, which showed no significant differences in the first six days ($p > 0.05$). In the AP group, the content reached 102.98 μg, which was slightly higher than those of the other two groups. However, the VP group was slightly higher than the MAP group on day 15. The change of the surface hydrophobicity of the protein may have been caused by the entry of nonpolar amino acid molecules into hydrophobic clusters [33]. The expansion or rearrangement of protein molecules could lead to a change of the secondary and tertiary structures of the proteins. VP and MAP significantly delayed the oxidation of protein in grouper meat, similar to what has been found in other fish studies [34]. The two-way analysis of variance showed that the storage time and packaging methods significantly affected the carbonyl content and hydrophobicity of the grouper fillets ($p < 0.01$). There was a significant difference between the VP and AP groups, and between the AP and MAP groups ($p < 0.01$); however, there was no significant difference between the VP and MAP groups ($p > 0.05$).

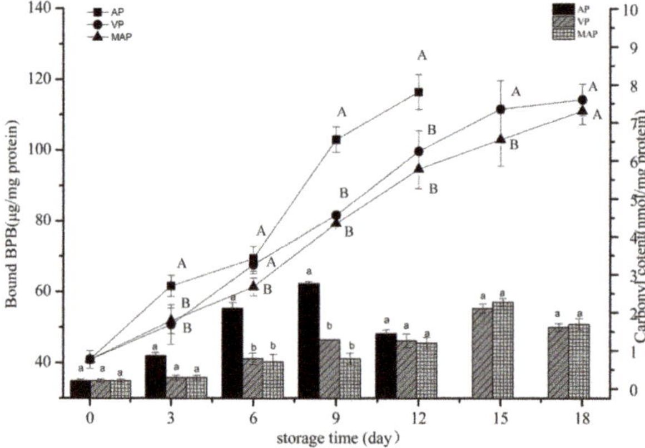

Figure 2. Changes in carbonyl content (bar graph) and protein surface hydrophobicity (expressed as Bound BPB, line graph) of grouper myofibrillar protein. (AP: air packaging; VP: vacuum packaging; MAP: modified atmosphere packaging; BPB: bromophenol blue).

3.3. Changes in Ca²⁺ ATPase Activity

Ca^{2+} ATPase mainly concentrates in the globular heads of myosin. The hydrophobic interactions, hydration of polar residues, and hydrogen bonds influence the stability of the three-dimensional structure of the protein. Since the three-dimensional structure of the protein determines the physiological activities of the protein itself, the activities of the protein may be lost or changed because of changes in the microstructure. The Ca^{2+} ATPase activity of actomyosin can be used as an important indicator for assessing the degree of protein denaturation, as it can indirectly reflect the integrity of myofibrillar protein [35].

Figure 3 shows that the Ca^{2+} ATPase activity of the grouper fillets respectively decreased in all samples. The two-way analysis of variance showed that storage time and packaging methods significantly affected the Ca^{2+} ATPase content of the grouper fillets ($p < 0.05$). On the 12th day, the content of the Ca^{2+} ATPase activity in the AP group decreased about 75.16%. There was a significant difference between the VP and AP groups and between the AP and MAP groups ($p < 0.01$), but there was no significant difference between the VP and MAP groups ($p > 0.05$). However, the VP and MAP groups were about 45.83%, which was consistent with the existing reports of other aquatic products [36]. It has been reported that the sulfhydryl group is abundant in the center of Ca^{2+} ATPase [21]. In our experiment, we found that the activity of Ca^{2+} ATPase was closely related to the sulfhydryl group. The correlation coefficients of the AP, VP, and MAP groups were 0.983, 0.946, and 0.984, respectively, and the Ca^{2+} ATPase activity was highly correlated with the change of the sulfhydryl group content. It was speculated that the oxidation of the sulfhydryl group in the active center resulted in a decrease of Ca^{2+} ATPase activity.

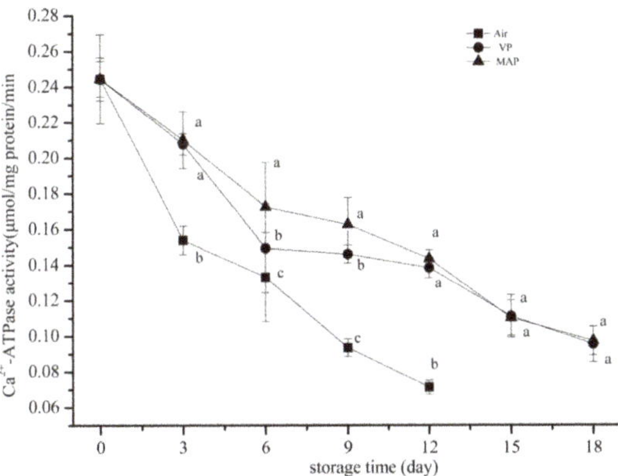

Figure 3. Changes in Ca^{2+} ATPase activity of grouper myofibrillar protein. (AP: air packaging; VP: vacuum packaging; MAP: modified atmosphere packaging).

3.4. Trichloroacetic Acid-Soluble Peptide Analysis

TCA-soluble peptides can reflect the degree of protein degradation. As shown in Figure 4, the TCA-soluble peptide content in all groups was obviously elevated during the entire storage period. The results showed that proteolysis occurred in succession and was similar to the trend reported by Yu et al. for grass carp fillets under refrigerated storage [37]. The initial number of TCA-soluble peptides of the grouper fillets was 0.42 µmol tyrosine/g sample, while the value reached 1.44 µmol tyrosine/g sample after six days storage for the AP group, which had the largest rate of increase among the three groups. The two-way analysis of variance revealed that the storage time and packaging methods significantly affected the TCA-soluble peptides of the grouper fillets ($p < 0.01$). The increase of TCA-soluble peptides might have initially been due to the activity of the endogenous enzyme [38]. Then, protein degradation was accelerated under the combined action of endogenous enzymes and microorganisms. The final TCA-soluble peptide content of the VP and MAP groups was significantly lower than that of the AP group ($p < 0.05$). The content of TCA-soluble peptides of the MAP group was lower than that of the VP group, but there was no significant difference between the two groups ($p < 0.05$), indicating the effective inhibition of proteolysis by VP and MAP, which might have been due to the growth of microorganisms and inhibition by the anoxic environment in the VP and MAP groups.

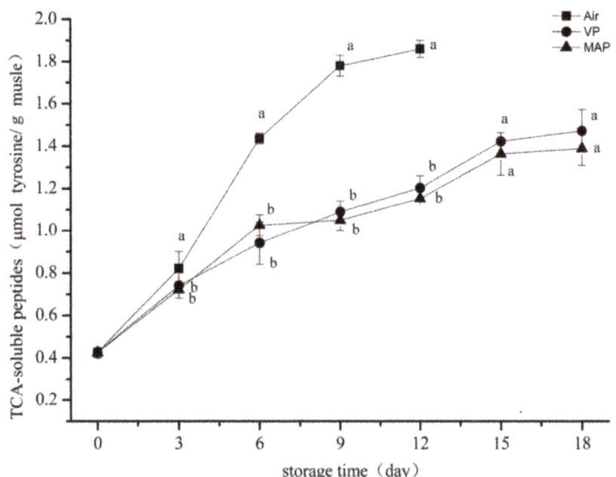

Figure 4. Changes in trichloroacetic acid (TCA)-soluble peptides of grouper myofibrillar protein (AP: air packaging; VP: vacuum packaging; MAP: modified atmosphere packaging).

3.5. Myofibril Fragmentation Index Analysis

As shown in Figure 5, the MFI maintained increasing trends among all groups. However, compared with the other two groups, the AP group rose significantly, with 97 mg/100 g being reached on day 3 ($p < 0.05$). No obvious change of the MFI in the VP and MAP groups was discovered at the corresponding time. The two-way analysis of variance revealed that storage time and packaging methods significantly affected the MFI of the grouper fillets ($p < 0.01$). There was no significant difference between the VP and MAP groups on the 12th day, but the MFI of the AP group was very significantly higher ($p < 0.01$) than those of the VP and MAP groups. This indicates that the internal integrity of myofibrillar in the AP group was the most destructive during refrigeration storage, while the destruction of the internal structure of myofibril was delayed by MAP and VP. Myofibril fragmentation refers to the phenomenon of myofibril breaking into shorter segments near the Z disk or Z line. The MFI was calculated from the percentage of myofibrils that were 1–4 sarcomeres long, which was mainly due to degradation of the connectin and actin of sarcomeric I. The MFI reflects the structural integrity of the myofibril during refrigerated storage [39].

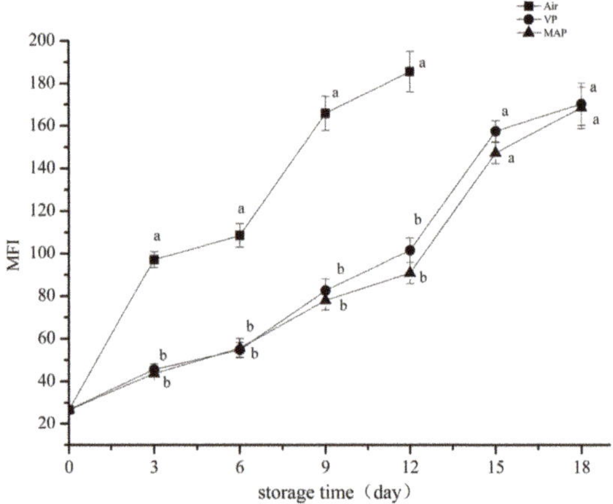

Figure 5. Changes in the myofibril fragmentation index (MFI) values of grouper myofibrillar protein. (AP: air packaging; VP: vacuum packaging; MAP: modified atmosphere packaging).

3.6. Free Amino Acid Analysis

The concentrations of FAAs in grouper fillets on the first, sixth, and twelfth days of refrigerated storage are shown in Table 1. The degradation of protein was the main reason for the variation of FAAs [37], so the degree of protein degradation could be reflected by the changes of the FAA content. A similar trend was also reported by Shi et al. for grass carp [40]. Table 1 shows that the content of FAA rapidly increased with storage time in the AP group, which might have been due to the degradation of protein caused by protein oxidation. However, the content of histidine in the grouper fillets was far below that found in grass carp, as reported by Yu et al., during refrigerated storage, which might be ascribed to the muscle composition of grouper. Aspartic acid, glutamic acid, and glycine are the main flavor contributors in the fillets, and the change of glycine was the most obvious among all groups. Glycine in the AP group increased from 91 mg/100 g on day 0 to 153 mg/100 g on day 6, then it dropped to 78 mg/100 g on day 12. The corresponding contents in the VP and MAP groups were 88 and 108 mg/100 g, respectively. The total FAA contents in the AP group were significantly higher ($p < 0.05$) compared with the other groups on day 6, and this phenomenon indicates that the degree of protein degradation was the greatest in the AP group. The number of FAAs was closely related to the storage time ($p < 0.05$), and the packaging method had no significant effect on it, as shown by the two-way analysis of variance ($p > 0.05$). The final FAA content was reduced among all groups, which when compared to day 6, might have been due to the degradation of enzymes caused by microbial growth [41].

Table 1. Changes in free amino acid (FAA) content (mg/100 g) of grouper muscle during refrigerated storage at 4 °C.

FAA	Day 0	AP Day 6	AP Day 12	VP Day 6	VP Day 12	MAP Day 6	MAP Day 12
Asp	5.69 ± 0.13 [a]	1.93 ± 0.12 [c]	1.09 ± 0.06 [c]	3.15 ± 0.12 [b]	1.71 ± 0.12 [c]	1.49 ± 0.05 [c]	3.32 ± 0.10 [b]
Thr	14.24 ± 0.48 [b]	20.57 ± 1.55 [a]	13.41 ± 1.10 [b,c]	20.84 ± 1.42 [a]	12.71 ± 0.14 [c]	12.65 ± 0.23 [c]	13.09 ± 0.15 [b,c]
Ser	20.24 ± 0.60 [b]	25.79 ± 0.16 [a]	13.78 ± 0.25 [b,c]	21.87 ± 1.57 [a]	12.25 ± 0.07 [c]	12.19 ± 0.11 [c]	14.01 ± 0.12 [b,c]
Glu	8.71 ± 0.30 [d]	27.20 ± 0.09 [a,b]	18.59 ± 0.31 [c]	18.34 ± 1.3 [c]	22.06 ± 0.1 [b,c]	30.98 ± 0.10 [a]	22.73 ± 0.01 [b,c]
Gly	91.14 ± 2.98 [b]	153.15 ± 1.21 [a]	78.74 ± 0.81 [c]	114.93 ± 7.94 [a,b]	88.33 ± 0.1 [b,c]	84.20 ± 0.85 [b,c]	108.33 ± 1.75 [a,b]
Ala	23.54 ± 0.78 [b,c]	43.80 ± 0.40 [a]	25.55 ± 0.25 [b,c]	27.65 ± 1.89 [b,c]	22.88 ± 0.04 [b,c]	30.04 ± 0.84 [b]	28.55 ± 0.49 [c]
Cys	1.23 ± 0.03 [a]	0.95 ± 0.11 [a,b]	0.48 ± 0.07 [c]	0.78 ± 0.09 [a,b]	0.24 ± 0.20 [d]	1.56 ± 0.09 [a]	0.48 ± 0.05 [c]
Val	3.69 ± 0.14 [c]	4.99 ± 0.12 [b,c]	3.90 ± 0.09 [c]	5.52 ± 0.38 [b]	2.91 ± 0.06 [c]	7.81 ± 0.04 [a]	5.77 ± 0.12 [b]
Met	1.70 ± 0.19 [d]	2.18 ± 0.20 [c]	1.90 ± 0.13 [d]	2.87 ± 0.11 [b,c]	1.16 ± 0.10 [d]	6.25 ± 0.11 [a]	3.25 ± 0.08 [b]
Ile	2.64 ± 0.15 [c]	3.43 ± 0.03 [b,c]	2.74 ± 0.11 [c]	3.89 ± 0.41 [b]	2.03 ± 0.13 [c]	5.47 ± 0.82 [a]	3.85 ± 0.03 [b]
Leu	4.18 ± 0.03 [c]	5.54 ± 0.09 [b]	4.44 ± 0.09 [c]	6.28 ± 0.49 [b]	3.12 ± 0.05 [c]	9.05 ± 0.03 [a]	6.39 ± 0.12 [b]
Tyr	1.61 ± 0.28 [c]	2.64 ± 0.26 [b,c]	1.55 ± 0.49 [c]	2.64 ± 0.37 [b,c]	1.62 ± 0.12 [c]	6.05 ± 0.85 [a]	2.93 ± 0.02 [b,c]
Phe	4.13 ± 0.39 [d]	5.34 ± 0.21 [c,d]	5.98 ± 0.14 [c]	6.37 ± 0.83 [b,c]	5.91 ± 0.39 [c]	9.54 ± 0.94 [a]	7.01 ± 0.42 [b,c]
Lys	33.66 ± 1.04 [b]	39.20 ± 0.29 [a,b]	30.85 ± 0.33 [c]	38.60 ± 2.87 [a,b]	36.00 ± 0.20 [b]	43.79 ± 0.27 [a]	28.30 ± 0.46 [c]
His	3.39 ± 0.25 [c]	5.27 ± 0.18 [a]	3.09 ± 0.91 [c]	4.74 ± 0.38 [b,c]	2.16 ± 0.10 [c,d]	4.61 ± 0.46 [b,c]	3.82 ± 0.34 [b,c]
Arg	4.57 ± 0.12 [c]	9.38 ± 0.48 [a]	7.61 ± 1.85 [a,b]	8.81 ± 0.49 [a]	5.82 ± 0.36 [b,c]	7.76 ± 0.37 [b,c]	5.24 ± 0.14 [a,b]
Pro	7.55 ± 0.26 [d]	10.67 ± 0.19 [b]	8.76 ± 0.15 [c,d]	17.42 ± 1.30 [a]	9.28 ± 0.07 [b,c]	18.78 ± 0.07 [a]	8.04 ± 0.37 [c,d]
total	231.91 ± 6.67 [c]	362.03 ± 4.03 [a]	219.49 ± 3.25 [c]	304.89 ± 21.5 [b]	230.19 ± 0.51 [c]	292.23 ± 6.03 [b,c]	265.09 ± 3.92 [b,c]

Results are expressed as means in mg per 100 g of sample with standard errors. Different lower-case letters ([a], [b], [c] and [d]) in different groups for same amino acid indicate a significant difference ($p < 0.05$). AP: air packaging group; VP: vacuum packaging group; MAP: modified atmosphere packaging group.

3.7. FTIR Analysis

The secondary structure of the protein was composed of an α-helix, a β-sheet, a β-turn, and a random coil. The function of a protein and its biochemical properties change with the variation of the secondary structure, which is due to the oxidation of the protein. As one of the main methods of studying the secondary structure of proteins, FTIR can be used to analyze changes in protein structure and the spatial distribution of proteins [42]. The amide I band of a protein (from 1600 to 1700 cm^{-1} of mid-infrared spectroscopy) can reveal a wealth of information about the constituents of the secondary structure in proteins. The peaks at the wavenumbers of 1600–1640, 1640–1650, 1650–1660, and 1660–1700 cm^{-1} are for the β-sheet, random coil, α-helix, and β-turn, respectively [43]. As shown in Figure 6, the second-order, second-derivative, mid-infrared spectra of the Gaussian fitting drawn by

PeakFit (PeakFit v 412, Systat Software Inc., USA) was used to analyze the protein secondary structure changes of each group at days 0 and 12.

Figure 6 shows that the peaks of each packaging group had a certain weakening on the 12th day. The change from the range of 1650–1660 cm^{-1} was significant compared with that of day 0, indicating that the levels of the random coil had increased. As shown in Figure 6A–D, the absorption peak of the spectrum shifted to the high-wavenumber area. The reason for this phenomenon was reported as the hydrogen bond of the protein structure being destroyed during refrigerated storage [44]. Since the spectrum diagram of the MAP group on the 12th day was most similar to that on day 0, compared to the other treatment groups, it indicates that the destruction of the secondary structure of the protein was inhibited in refrigerated grouper fillets in the MAP group.

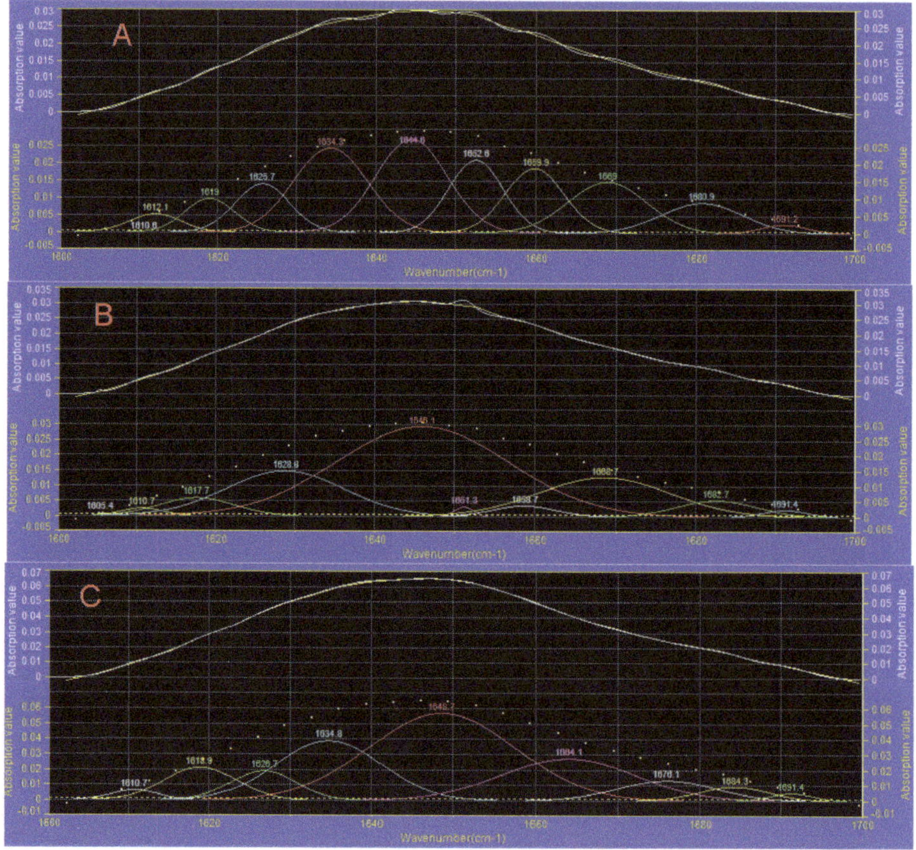

Figure 6. *Cont.*

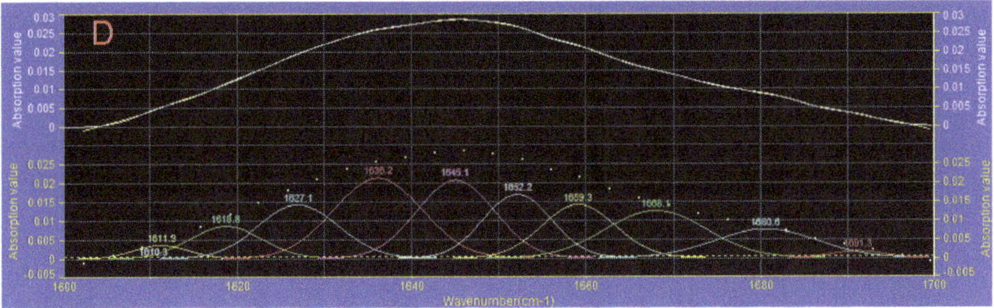

Figure 6. The second-order, second-derivative, mid-infrared spectra by PeakFit. (**A**: 0 days; **B**: AP group on the 12th day; **C**: VP group on the 12th day; **D**: MAP group on the 12th day).

The change in the secondary structure for all groups is shown in Figure 7. The random coil increased, and the α-helix content decreased during refrigerated storage, which might have been due to the breakage of the hydrogen bonds. Further, the surface hydrophobicity increased, and a disordered state gradually became present in the protein. The random coil increased by 34.8% on the 12th day compared to day 0 in the AP group, which was significantly higher than that of the VP and MAP groups ($p > 0.05$). The results showed that the protein structure was partly inhibited by VP and MAP. The β-sheet of each group also increased to a certain extent ($p > 0.05$), due to the change of the peptide chain folding structure caused by the gradual formation of sulfhydryl oxidation and disulfide bonds.

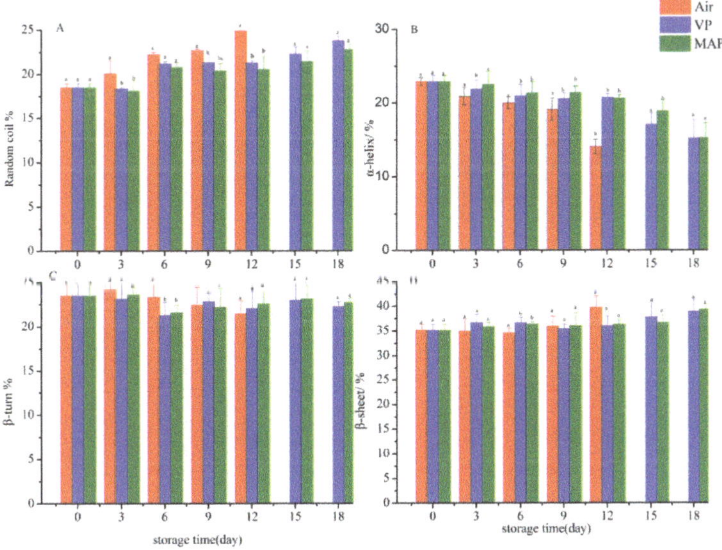

Figure 7. Changes in secondary structure contents of grouper protein during refrigerated storage (**A**: the change in the random coil content under different treatments, **B**: the change in the α-helix content under different treatments, **C**: the change in the β-turn content under different treatments, **D**: the change in the β-sheet content under different treatments). (Air: air packaging; VP: vacuum packaging; MAP: modified atmosphere packaging).

3.8. SDS-PAGE Analysis

The total protein electrophoresis results of the different treatment groups are shown in Figure 8. The results of the band optical density analysis using Quantity One 4.0 showed that the content of the myosin heavy chain (MHC) and actin increased between 0 to 6 days in different groups. Then, different degrees of degradation appeared in different groups after six days. At day 12, the intensity of the MHC in the AP, VP, and MAP groups was 355, 555, and 643, respectively ($p < 0.05$). The MHC, which is very easily oxidized, is crosslinked by disulfide and non-disulfide covalent bonds which contribute to the formation of high-molecular-weight polymers and aggregates [45]. The degradation of MHC was inhibited significantly by VP and MAP. The degradation of protein was probably due to MHC oxidation, and oxidative damage of proteins may have also led to protein degradation and the crosslinking and aggregation of actin, which corresponds to the results of Lu et al. [46]. The actomyosin (42 kDa) band gradually became shallow over time. The change ranges of the VP and MAP groups were significantly smaller than that of the AP group, the main reason for which being the hydrolysis of cathepsin L; the change range of the MAP group was the smallest, indicating that the activity of cathepsin L in the MAP group was inhibited. Band III (13 kDa) of the AP group at day 12 was significantly different than the other two groups, and was highly correlated with reduced actin (correlation coefficient of 0.89), which we concluded was due to the degradation product of actin.

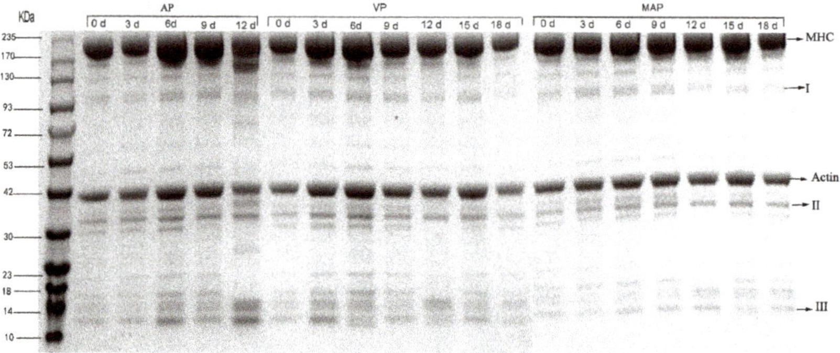

Figure 8. Effects of three packaging methods (AP, VP, and MAP) on total protein degradation of grouper samples during refrigerated storage at 4 °C. (AP: air packaging; VP: vacuum packaging; MAP: modified atmosphere packaging).

It has not been reported in previous works that all of the intensities of band II (34 kDa) in the three groups showed an increasing trend in the first 12 days. It can be concluded that the intensity of band II, which was speculated to be an indicator protein of grouper freshness before the 12th day, had a highly linear relationship with storage time ($R^2 = 0.97$), as analyzed by Quantity One 4.0. However, in the VP and MAP groups, the intensity of band II decreased after 12 days. The possible reason for this is that the protein belongs to a water-soluble protein, and the rate of drip loss increases at the later stage of storage [47]. The concentration of protein decreased with water loss, due to the type of protein belonging to water-soluble proteins; such changes can also be observed in bands I and III.

This study showed that the increase of protein oxidation was reflected by the content and distribution of carbonyl and sulfhydryl content changes. Table 2 shows that the carbonyl and sulfhydryl contents of protein oxidation indexes in each group were highly correlated with protein degradation, TCA-soluble peptides, and MFI. This indicates that protein oxidation promotes the degradation process during refrigerated storage of grouper fillets. Compared with AP, VP and MAP exhibited significant positive effects with regard to reducing protein oxidation. However, the MAP group showed better effectiveness than the VP group, which may have been due to the higher residual oxygen and oxygen transmission rate in VP. When vacuum pumping cannot make a complete vacuum environment, which results in residual air, the external oxygen gradually permeates into the packaging under the effect of internal and external pressure differences.

It is also speculated that a high concentration of carbon dioxide not only controls the growth of microorganisms, but also changes the pH on the surface of the fish, thus changing the activity of protease and affecting the degradation of myofibrils.

Unlike the results of Lametsch and Lonergan, which showed that protein oxidation reduces the activity of μ-calpain cysteine [48], thus inhibiting the degradation of muscle fibril in pork, the reason here may be that the low rate of protein oxidation could enhance the sensitivity of proteolytic enzymes for myofibril [14], and the activity of protease loss is not obvious under weak oxidation. Thus, fish and other meat proteasome systems also show large differences, which also might be the reason for the difference between the experimental results. Therefore, further research is needed on the metabolic pathway of how the protein oxidation of grouper muscle affects protein degradation.

Table 2. Correlation analysis between protein oxidation and the degradation of grouper in three groups.

Group	Indicator of Protein Oxidation	Indicator of Protein Degradation	Pearson Correlation Coefficient
AP	Carbonyl content	TCA-soluble peptide content	0.683
AP	Total sulfhydryl content	TCA-soluble peptide content	−0.981 **
AP	Carbonyl content	Myofibril fragmentation index	0.956 *
AP	Total sulfhydryl content	Myofibril fragmentation index	−0.961 *
VP	Carbonyl content	TCA-soluble peptide content	0.941 *
VP	Total sulfhydryl content	TCA-soluble peptide content	−0.975 *
VP	Carbonyl content	Myofibril fragmentation index	0.940 *
VP	Total sulfhydryl content	Myofibril fragmentation index	−0.975 **
MAP	Carbonyl content	TCA-soluble peptide content	0.793
MAP	Total sulfhydryl content	TCA-soluble peptide content	−0.982 **
MAP	Carbonyl content	Myofibril fragmentation index	0.934 *
MAP	Total sulfhydryl content	Myofibril fragmentation index	−0.929 *

AP: air packaging; VP: vacuum packaging; MAP: modified atmosphere packaging. * correlation ** strong correlation.

4. Conclusions

This study shows that the changes of carbonyl, sulfhydryl, and Ca^{2+} ATPase activity greatly varied with prolonged storage time, and demonstrates that the degree of grouper fillet protein oxidation was increased. The amide I band absorption peak of the infrared moved towards a higher wavenumber, while the secondary structure of the α-helix gradually transformed into a random curl. It has been shown that great changes of protein structure took place in grouper fillets during refrigerated storage. By combining indexes of protein degradation, such as MFI, SDS-PAGE, and TCA-soluble peptide content, it is concluded that myofibril oxidation could promote protein degradation in grouper fillets during refrigerated storage, which could be observed in each group (AP, VP, and MAP). High-carbon-dioxide MAP played a positive role in the inhibition of myofibril degradation and oxidation for refrigerated grouper fillets.

Author Contributions: Conceptualization, X.Z. and J.X.; methodology, J.X.; software, X.Z.; validation, J.X.; data curation, X.Z.; writing—original draft preparation, X.Z.; writing—review and editing, W.H.; project administration, W.H.; funding acquisition, J.X.

Funding: This research was financially supported by National Natural Science Foundation of China (grant number: 31571914), the China Agriculture Research System (CARS-47), the Construction Project of the Public Service Platform for Shanghai Municipal Science and Technology Commission (17DZ2293400), and the Shanghai Municipal Science and Technology Project to enhance the capabilities of the platform (grant number: 19DZ2284000).

Conflicts of Interest: The authors declare no conflict of interest.

References

1. Kirtil, E.; Kilercioglu, M.; Oztop, M.H. Modified atmosphere packaging of foods. In *Reference Module in Food Science*; Elsevier: Amsterdam, The Netherlands, 2016.

2. Guimarães, C.F.M.; Mársico, E.T.; Monteiro, M.L.G.; Lemos, M.; Mano, S.B.; Junior, C.A.C. The chemical quality of frozen vietnamese pangasius hypophthalmus fillets. *Food Sci. Nutr.* **2016**, *4*, 398–408. [CrossRef] [PubMed]

3. Zhao, H.; Liu, S.; Tian, C.; Yan, G.; Wang, D. An overview of current status of cold chain in china. *Int. J. Refrig.* **2018**, *88*, 483–495. [CrossRef]

4. Estévez, M. Protein carbonyls in meat systems: A review. *Meat Sci.* **2011**, *89*, 259–279. [CrossRef] [PubMed]

5. Turgut, S.S.; Işıkçı, F.; Soyer, A. Antioxidant activity of pomegranate peel extract on lipid and protein oxidation in beef meatballs during frozen storage. *Meat Sci.* **2017**, *129*, 111–119. [CrossRef] [PubMed]

6. Ge, L.; Xu, Y.; Xia, W.; Jiang, Q.; Jiang, X. Differential role of endogenous cathepsin and microorganism in texture softening of ice-stored grass carp (ctenopharyngodon idella) fillets. *J. Sci. Food Agric.* **2016**, *96*, 3233–3239. [CrossRef] [PubMed]

7. Lu, H.; Liu, X.; Zhang, Y.; Wang, H.; Luo, Y. Effects of chilling and partial freezing on rigor mortis changes of bighead carp (aristichthys nobilis) fillets: Cathepsin activity, protein degradation and microstructure of myofibrils. *J. Food Sci.* **2016**, *80*, C2725–C2731. [CrossRef] [PubMed]

8. Zhang, W.; Xiao, S.; Ahn, D.U. Protein oxidation: Basic principles and implications for meat quality. *Crit. Rev. Food Sci. Nutr.* **2013**, *53*, 1191–1201. [CrossRef]

9. Van Haute, S.; Raes, K.; Devlieghere, F.; Sampers, I. Combined use of cinnamon essential oil and map/vacuum packaging to increase the microbial and sensorial shelf life of lean pork and salmon. *Food Packag. Shelf Life* **2017**, *12*, 51–58. [CrossRef]

10. dos Santos, P.R.; Donado-Pestana, C.M.; Delgado, E.F.; Tanaka, F.O.; Contreras-Castillo, C.J. Tenderness and oxidative stability of nellore bulls steaks packaged under vacuum or modified atmosphere during storage at 2 °C. *Food Packag. Shelf Life* **2015**, *4*, 10–18. [CrossRef]

11. Łopacka, J.; Półtorak, A.; Wierzbicka, A. Effect of map, vacuum skin-pack and combined packaging methods on physicochemical properties of beef steaks stored up to 12days. *Meat Sci.* **2016**, *119*, 147–153. [CrossRef]

12. Aguilera Barraza, F.A.; León, R.A.Q.; Álvarez, P.X.L. Kinetics of protein and textural changes in atlantic salmon under frozen storage. *Food Chem.* **2015**, *182*, 120–127. [CrossRef] [PubMed]

13. Pan, S.; Wu, S. Effect of chitooligosaccharides on the denaturation of weever myofibrillar protein during frozen storage. *Int. J. Biol. Macromol.* **2014**, *65*, 549–552. [CrossRef] [PubMed]

14. Yang, F.; Jing, D.; Yu, D.; Xia, W.; Jiang, Q.; Xu, Y.; Yu, P. Differential roles of ice crystal, endogenous proteolytic activities and oxidation in softening of obscure pufferfish (takifugu obscurus) fillets during frozen storage. *Food Chem.* **2019**, *278*, 452–459. [CrossRef] [PubMed]

15. Davies, K.J.A. Degradation of oxidized proteins by the 20 s proteasome. *Biochimie* **2001**, *83*, 301–310. [CrossRef]

16. Berardo, A.; Claeys, E.; Vossen, E.; Leroy, F.; De Smet, S. Protein oxidation affects proteolysis in a meat model system. *Meat Sci.* **2015**, *106*, 78–84. [CrossRef] [PubMed]

17. Ogawa, M.; Nakamura, S.; Horimoto, Y.; An, H.; Tsuchiya, T.; Nakai, S. Raman spectroscopic study of changes in fish actomyosin during setting. *J. Agric. Food Chem.* **1999**, *47*, 3309. [CrossRef] [PubMed]

18. Abbey, L.D.; Glover-Amengor, M.; Atikpo, M.O.; Howell, N.K. Proximate and biochemical characterization of burrito (bachydeuterus auritus) and flying gurnard (dactylopterus volitans). *Food Sci. Nutr.* **2017**, *5*, 369–373. [CrossRef] [PubMed]

19. Oliver, C.N.; Ahn, B.W.; Moerman, E.J.; Goldstein, S.; Stadtman, E.R. Age-related changes in oxidized proteins. *J. Biol. Chem.* **1987**, *262*, 5488.

20. Chelh, I.; Gatellier, P.; Santé-Lhoutellier, V. Technical note: A simplified procedure for myofibril hydrophobicity determination. *Meat Sci.* **2006**, *74*, 681–683. [CrossRef]

21. Benjakul, S.; Visessanguan, W.C.; Tanaka, M. Comparative study on physicochemical changes of muscle proteins from some tropical fish during frozen storage. *Food Res. Int.* **2003**, *36*, 787–795. [CrossRef]

22. Thannhauser, T.W.; Konishi, Y.; Scheraga, H.A. Analysis for disulfide bonds in peptides and proteins. *Methods Enzymol.* **1987**, *143*, 115. [PubMed]

23. Benjakul, S.; Seymour, T.A.; Morrissey, M.T.; Haejung, A.N. Physicochemical changes in pacific whiting muscle proteins during iced storage. *J. Food Sci.* **2010**, *62*, 729–733. [CrossRef]

24. Thanonkaew, A.; Benjakul, S.; Visessanguan, W.; Decker, E.A. The effect of metal ions on lipid oxidation, colour and physicochemical properties of cuttlefish (sepia pharaonis) subjected to multiple freeze–thaw cycles. *Food Chem.* **2006**, *95*, 591–599. [CrossRef]

25. Sriket, C.; Benjakul, S.; Visessanguan, W.; Hara, K.; Yoshida, A. Retardation of post-mortem changes of freshwater prawn (macrobrachium rosenbergii) stored in ice by legume seed extracts. *Food Chem.* **2012**, *135*, 571–579. [CrossRef] [PubMed]

26. Culler, R.D.; Parrish, F.C., Jr.; Smith, G.C.; Cross, H.R. Relationship of myofibril fragmentation index to certain chemical, physical and sensory characteristics of bovine longissimus muscle. *J. Food Sci.* **2010**, *43*, 1177–1180. [CrossRef]

27. Yu, D.; Xu, Y.; Regenstein, J.M.; Xia, W.; Yang, F.; Jiang, Q.; Wang, B. The effects of edible chitosan-based coatings on flavor quality of raw grass carp (ctenopharyngodon idellus) fillets during refrigerated storage. *Food Chem.* **2018**, *242*, 412–420. [CrossRef] [PubMed]

28. Verrez-Bagnis, V.; Ladrat, C.; Morzel, M.; Noël, J.; Fleurence, J. Protein changes in post mortem sea bass (dicentrarchus labrax) muscle monitored by one-and two-dimensional gel electrophoresis. *Electrophoresis* **2015**, *22*, 1539–1544. [CrossRef]

29. Biswas, S.; Chida, A.S.; Rahman, I. Redox modifications of protein-thiols: Emerging roles in cell signaling. *Biochem. Pharmacol.* **2006**, *71*, 551–564. [CrossRef]

30. Soyer, A.; Özalp, B.; Dalmış, Ü.; Bilgin, V. Effects of freezing temperature and duration of frozen storage on lipid and protein oxidation in chicken meat. *Food Chem.* **2010**, *120*, 1025–1030. [CrossRef]

31. Nikoo, M.; Benjakul, S.; Rahmanifarah, K. Hydrolysates from marine sources as cryoprotective substances in seafoods and seafood products. *Trends Food Sci. Technol.* **2016**, *57*, 40–51. [CrossRef]

32. Zhang, L.; Gui, P.; Zhang, Y.; Lin, J.; Li, Q.; Hong, H.; Luo, Y. Assessment of structural, textural, and gelation properties of myofibrillar protein of silver carp (hypophthalmichthys molitrix) modified by stunning and oxidative stress. *LWT-Food Sci. Technol.* **2019**, *102*, 142–149. [CrossRef]

33. Yarnpakdee, S.; Benjakul, S.; Visessanguan, W.; Kijroongrojana, K. Thermal properties and heat-induced aggregation of natural actomyosin extracted from goatfish (mulloidichthys martinicus) muscle as influenced by iced storage. *Food Hydrocoll.* **2009**, *23*, 1779–1784. [CrossRef]

34. He, Y.; Huang, H.; Li, L.; Yang, X.; Hao, S.; Chen, S.; Deng, J. The effects of modified atmosphere packaging and enzyme inhibitors on protein oxidation of tilapia muscle during iced storage. *LWT-Food Sci. Technol.* **2018**, *87*, 186–193. [CrossRef]

35. Ko, W.-C.; Shi, H.-Z.; Chang, C.-K.; Huang, Y.-H.; Chen, Y.-A.; Hsieh, C.-W. Effect of adjustable parallel high voltage on biochemical indicators and actomyosin Ca^{2+}-atpase from tilapia (*Orechromis niloticus*). *LWT-Food Sci. Technol.* **2016**, *69*, 417–423. [CrossRef]

36. Zhang, B.; Hao, G.-J.; Cao, H.-J.; Tang, H.; Zhang, Y.-Y.; Deng, S.-G. The cryoprotectant effect of xylooligosaccharides on denaturation of peeled shrimp (litopenaeus vannamei) protein during frozen storage. *Food Hydrocoll.* **2018**, *77*, 228–237. [CrossRef]

37. Yu, D.; Regenstein, J.M.; Zang, J.; Xia, W.; Xu, Y.; Jiang, Q.; Yang, F. Inhibitory effects of chitosan based coatings on endogenous enzyme activities, proteolytic degradation and texture softening of grass carp (ctenopharyngodon idellus) fillets stored at 4 °C. *Food Chem.* **2018**, *262*, 1. [CrossRef] [PubMed]

38. Rawdkuen, S.; Jongjareonrak, A.; Phatcharat, S.; Benjakul, S. Assessment of protein changes in farmed giant catfish (pangasianodon gigas) muscles during refrigerated storage. *Int. J. Food Sci. Technol.* **2010**, *45*, 985–994. [CrossRef]

39. Taylor, R.G.; Geesink, G.H.; Thompson, V.F.; Koohmaraie, M.; Goll, D.E. Is z-disk degradation responsible for postmortem tenderization? *J. Anim. Sci.* **1995**, *73*, 1351. [CrossRef] [PubMed]

40. Shi, C.; Cui, J.; Na, Q.; Luo, Y.; Han, L.; Hang, W. Effect of ginger extract and vinegar on atp metabolites, imp-related enzyme activity, reducing sugars and phosphorylated sugars in silver carp during postslaughter storage. *Int. J. Food Sci. Technol.* **2017**, *52*, 413–423. [CrossRef]

41. Ruiz-Capillas, C.; Moral, A. Changes in free amino acids during chilled storage of hake (merluccius merluccius L.) in controlled atmospheres and their use as a quality control index. *Eur. Food Res. Technol.* **2001**, *212*, 302–307. [CrossRef]

42. Hanne, C.B.; Achim, K.; Ulrike, B.C.; Ragni, O.; Andersen, H.J. Heat-induced changes in myofibrillar protein structures and myowater of two pork qualities. A combined ft-ir spectroscopy and low-field nmr relaxometry study. *J. Agric. Food Chem.* **2006**, *54*, 1740–1746.

43. Carton, I.; Bocker, U.; Ofstad, R. Monitoring secondary structural changes in salted and smoked salmon muscle myofiber proteins by ft-ir microspectroscopy. *J. Agric. Food Chem.* **2009**, *57*, 3563–3570. [CrossRef] [PubMed]

44. Tu, Z.C.; Huang, T.; Wang, H.; Sha, X.M.; Shi, Y.; Huang, X.Q.; Man, Z.Z.; Li, D.J. Physico-chemical properties of gelatin from bighead carp (hypophthalmichthys nobilis) scales by ultrasound-assisted extraction. *J. Food Sci. Technol.* **2015**, *52*, 2166–2174. [CrossRef] [PubMed]

45. Xiong, Y.L.; Blanchard, S.P.; Tooru, O.; Yuanyuan, M. Hydroxyl radical and ferryl-generating systems promote gel network formation of myofibrillar protein. *J. Food Sci.* **2010**, *75*, C215–C221. [CrossRef] [PubMed]

46. Lu, H.; Zhang, L.; Li, Q.; Luo, Y. Comparison of gel properties and biochemical characteristics of myofibrillar protein from bighead carp (aristichthys nobilis) affected by frozen storage and a hydroxyl radical-generation oxidizing system. *Food Chem.* **2017**, *223*, 96–103. [CrossRef] [PubMed]

47. Cayuela, J.M.; Gil, M.D.; Bañón, S.; Garrido, M.D. Effect of vacuum and modified atmosphere packaging on the quality of pork loin. *Eur. Food Res. Technol.* **2004**, *219*, 316–320. [CrossRef]

48. Lametsch, R.; Lonergan, S.; Huff-Lonergan, E. Disulfide bond within mu-calpain active site inhibits activity and autolysis. *Biochim. Et Biophys. Acta Proteins Proteom.* **2008**, *1784*, 1215–1221. [CrossRef]

 *foods*

Article

Stabilization of Crystalline Carotenoids in Carrot Concentrate Powders: Effects of Drying Technology, Carrier Material, and Antioxidants

Klara Haas [1],*, **Paul Robben** [2], **Anke Kiesslich** [2], **Marcus Volkert** [2] and **Henry Jaeger** [1]

[1] Department of Food Science and Technology, University of Natural Resources and Life Sciences (BOKU), 1190 Vienna, Austria
[2] GNT Europa GmbH, 52072 Aachen, Germany
* Correspondence: Klara.haas@boku.ac.at

Received: 29 June 2019; Accepted: 23 July 2019; Published: 25 July 2019

Abstract: Coloring concentrates of carotenoid-rich plant materials are currently used in the food industry to meet the consumer's demand for natural substitutes for food colorants. The production of shelf-stable powders of such concentrates comes with particular challenges linked to the sensitivity of the active component towards oxidation and the complexity of the composition and microstructure of such concentrates. In this study, different strategies for the stabilization of crystalline carotenoids as part of a natural carrot concentrate matrix during drying and storage were investigated. The evaluated approaches included spray- and freeze drying, the addition of functional additives, and oxygen free storage. Functional additives comprised carrier material (maltodextrin, gum Arabic, and octenyl succinic anhydride (OSA)-modified starch) and antioxidants (mixed tocopherols, sodium ascorbate). Degradation and changes in the physical state of the carotenoid crystals were monitored during processing and storage. Carotenoid losses during processing were low (>5%) irrespective of the used technology and additives. During storage, samples stored in nitrogen showed the highest carotenoid retention (97–100%). The carotenoid retention in powders stored with air access varied between 12.3% ± 2.1% and 66.0% ± 5.4%, having been affected by the particle structure as well as the formulation components used. The comparative evaluation of the tested strategies allows a more targeted design of processing and formulation of functional carrot concentrate powders.

Keywords: spray drying; freeze drying; antioxidants; carotenoid aggregates; coloring foods

---

1. Introduction

With consumer demands shifting towards more natural ingredients and clean labeling, functional vegetable concentrates containing a high carotenoid content present an attractive coloring alternative to natural and artificial color additives. Compared to the use of isolated carotenoids as food colorants, the advantage of "coloring foods" such as carrot concentrates, lies in the recovery without a solvent dependent, selective extraction step of the main coloring components, which is often perceived as more natural by consumers.

However, the replacement of artificial colors with coloring vegetable concentrates comes with certain challenges, which are linked to the oxidative susceptibility and the crystalloid nature of the carotenoids as active ingredient as well as the complex matrix containing multiple carrot derived co-components [1,2]. Carrot carotenoids (e.g., β-carotene, α-carotene, lutein), are lipophilic pigments which tend to form supramolecular aggregates in hydrophilic environments [3]. In raw carrots (*Daucus carota*), most of the carotenoids are present in a crystalline state derived from accumulated carotenoid aggregation in carrot chromoplasts [4]. The presence of crystalline carotenoids as well as a small fraction of monomolecular (dissolved) carotenoids is responsible for the typical orange hue of the carrot, whereas a yellow hue occurs when the carotenoid pigments are completely dissolved in organic solvents or lipids [5]. Thus, when color is a target function of carrot concentrate powders, both the monitoring of this supramolecular conformation as well as the total pigment concentration during processing and storage is of importance.

In some of their physical properties crystalline carotenoids differ from the lipid-dissolved carotenoids and have also shown to react differently to certain processing conditions such as heat [3,6,7]. While emulsions, in which carotenoids are dissolved in oil droplets, are suitable delivery systems for carotenoids, the emulsification process mostly requires dissolving crystals, upon which the characteristics (particulate character and hue) of the crystalline carotenoids are lost. For the preservation of the desired properties, processing and drying requires an approach with a minimal effect upon the crystalline structure. Understanding the impact of formulation and powder processing on the crystalline structure of the carotenoids, as well as their degradation during processing and storage is crucial in order to provide high quality functional food systems. Only limited information on the mechanism and the role of different impact factors for the stabilization of carotenoid in powders made from complex vegetable concentrates is available and the relevance of the crystallinity has never been studied in this context. Additionally, the analysis and identification of crystalline carotenoids within the complex matrix of plant concentrates still presents an analytical challenge, since their concentration within the concentrates is too low to apply methodology which can be used for the determination of crystallinity of pure and highly concentrated components (e.g., X-ray diffraction analysis and differential scanning calorimetry).

Due to a change in the optical properties of carotenoids when they change from the dissolved to the aggregated or crystalline state, their macromolecular conformation can be determined by optical methods such as UV/Vis spectroscopy and circular dichroism spectroscopy [8]. The measurement of the crystalline carotenoids in a hydrophilic dilution is hindered by the high turbidity of the carrot concentrate dispersion. On the other hand, carotenoid extraction through organic solvents, or concentration which is necessary for most carotenoid quantification methods, leads to a crystal dissolution, which prevents a characterization of the physical state. Hence, suitable analytical procedures have to be developed to derive information about the macromolecular conformation and its changes through processing within the turbid sample matrix.

The present study investigated the stability of carotenoid crystals as part of a natural carrot matrix during the preparation of spray dried (SD) and freeze dried (FD) powders and during further storage. Different process and recipe-based approaches to increase the shelf life of functional carrot concentrate powders were evaluated with regard to their effect on carotenoid stability and integrity of their crystalline state. Whereas the general impact of the investigated approaches is already known for many monomolecular active ingredients, the aim of this study was the quantification of the effect of aforementioned factors for the given complex food system containing crystalline carotenoids. Since the shelf life of carrot concentrates is strongly linked to the preservation of the hue provided by the crystalline state of the carotenoids as well as by their concentration, both the stability of the physical state and the pigment concentration during production and storage were monitored. An analytical approach was applied which allowed for the qualitative assessment of the physical state of the carotenoids based on their UV/Vis spectrum. Thereby, it was possible to specifically monitor the impact of the different formulation strategies on the stability of the macromolecular conformation.

2. Materials and Methods

2.1. Materials

Orange carrot concentrate (69 °Bx), provided by GNT Group B. V. (Mierlo, The Netherlands) was used as a basis material for the production of carrot concentrate powders. Maltodextrin (MD) with a dextrose equivalent (DE) 12 (Glucidex IP12, Roquette, Lestrem, France) was used as a standard carrier for all trials when not otherwise stated. The antioxidants being tested were mixed tocopherols (Toc) (L-70 IP, Jan Dekker, Amsterdam, The Netherlands) and sodium ascorbate (SA) (Roth, Karlsruhe, Germany,). According to the manufacturer the mixed tocopherols contained D-beta-, delta-, alpha-, and gamma-tocopherol and sunflower oil (<30% *w/w*). The surface-active carriers used were octenyl succinic anhydride (OSA)-modified starch (C*emcap 12635, Cargill, Redon, France) and gum Arabic (GA) (Seyal, Encapsia, Nexira, Rouen, France).

2.2. Tested Recipe Parameters

Tested recipe parameters and their concentrations are shown in Table 1. Reference samples (SD-Ref) were produced with the carrot concentrate at the beginning, during, and at the end of sample production with MD as the sole carrier. Samples with surface-active carrier material were produced by substituting 50% and 100% (*w/w*) of the MD in the recipe with the respective carrier. To evaluate the effect of antioxidant addition (AO), a hydrophilic (sodium ascorbate) and lipophilic (mixed tocopherols) antioxidant was chosen. The antioxidants were added in low concentrations (250–500 µg/g) or high concentrations (2500–5000 µg/g), whereas the concentration refers to the theoretical concentration in the dry matter. The high concentrations tested in this study were derived from preliminary trials in which they showed no pro-oxidant effect.

Table 1. Concentration of tested recipe components in the produced carrot concentrate powders. Residue dry matter (DM) consists of carrot concentrate constituents.

Sample Code	Toc [1]	SA [2]	MD [3]	OSA [4]	GA [5]
	(µg/g DM)	(µg/g DM)	(g/g DM)	(g/g DM)	(g/g DM)
SD-Ref [a]/FD [b]/SD-no-HPH [b]/SD-E [a]	-	-	0.5	-	-
SD 50% GA [b]	-	-	0.25	-	0.25
SD 100% GA [b]	-	-	-	-	0.5
SD 50% OSA [b]	-	-	0.25	0.25	-
SD 100% OSA [b]	-	-	-	-	0.5
SD-Toc-low [c]	250	-	0.5	-	-
SD-SA-low [c]		500	0.5	-	-
SD-Toc-high [b]	2500	-	0.5	-	-
SD-SA-high [b]		5000	0.5	-	-
SD-Toc-SA-low [b]	250	500	0.5	-	-
SD-Toc-SA-high [c]	2500	5000	0.49	-	-

[1] mixed tocopherols, [2] sodium ascorbate, [3] maltodextrin DE 12, [4] OSA-starch, [5] gum Arabic, Letters indicate the amount of process replicates (n) for the trial: [a]: $n = 3$; [b]: $n = 2$; [c]: $n = 1$. SD: spray dried; FD: freeze dried; SD-Ref: spray dried reference; SD-no-HPH: samples without high-pressure homogenization; SD-E: spray dried powder collected from the drying chamber.

2.3. Slurry Production

From the carrot concentrate, a batch of 2.5 kg slurry with a total dry matter content of 35% (*w/w*) was prepared for each trial. To avoid structural collapse due to humidity caking during storage, the dry matter of the carrier material accounted for 50% of the slurry dry matter. To ensure complete hydration, solutions of OSA-starch and GA were prepared 4 h prior to being mixed with the concentrate and other ingredients. Distilled water was heated up to 70–85 °C and mixed with the carrier and concentrate by means of a rotor-stator dispersing unit (Ultra-Turrax 50, IKA, Staufen, Germany) at 2000 rpm until the slurry was visually homogenous (5–7 min). Each batch was further homogenized for 10 min at 7000 rpm to ensure complete dissolution and homogenous distribution of the recipe components. Toc were added during slurry preparation along with the carrier material, while the more heat sensitive SA was stirred in by hand after 10 min of homogenization. Except for specific samples which were produced without high-pressure homogenization (SD-no-HPH), the slurries were further subjected to high-pressure homogenization in a two-stage homogenizer (Gaulin, APV, Luebeck, Germany) at 35/7 MPa.

2.4. Spray Drying and Freeze Drying

Freshly prepared slurries were stirred on a magnetic-stirrer and spray dried (SD) using a pilot scale spray dryer (Anhydro, Søborg, Denmark) equipped with a two-fluid nozzle. The spray drying was carried out in a co-current mode and conditions were kept constant in all trials at 195 °C inlet air temperature and 80 °C outlet air temperature. The atomizing pressure was set at 0.15 MPa, the feed flow rate was 40 g/min, and an amount of 1.5 kg of each batch was dried in one run. Produced powders were collected in a sampling container attached to a cyclone separator. For selected trials, an additional fraction (SD-E), consisting of dried particles with a larger median particle size, was recovered from the drying tower after completion of the process by gently sweeping the wall of the drying tower with a brush and subsequently collecting the thus attained powder. Freeze drying was performed to compare spray drying to a low temperature drying process. Slurries were poured into plastic containers to a total high of 8 mm and frozen in a shock freezer to −40 °C. The frozen slurries were then kept at −30 °C until freeze drying in a laboratory freeze dryer (Labconco Coop, Kansas City, MO, USA) at 0.0133 MPa for 5 days. After freeze drying, powders were milled in a blender (Nutribullet, Trencin, Slovakia) to a particle size below 200 µm.

2.5. Dry Matter Content

The dry matter content (DM) of the powders and the concentrate was determined gravimetrically. Powder samples of 3 g were dried for 24 h at 89 °C and the DM was calculated as ratio of the sample weight before drying and after drying. Concentrates and slurries were dispersed in dried sand to increase the total surface area before drying.

2.6. Assessment of the Physical State of Carotenoids

In order to measure the UV/Vis absorbance spectra of the unextracted carrot crystals, concentrate and powder samples were dissolved in water and further diluted in a hydrophilic medium. The dilution of samples with a concentrated sugar solution (69 °Bx, Grafschaft Krautfabrik, Meckenheim, Germany), successfully led to significantly reduced turbidity and light scattering interference compared to samples diluted with water. Figure 1 shows the UV/Vis spectra of the carrot concentrate diluted with water (A), sugar solution and water (B, 1:1 *v/v*), and pure sugar solution (C). Carrot concentrate diluted in sugar solution displayed a pronounced reduction of matrix interference while providing a reproducible measurement, which could then be used to derive further information concerning the physical state of the carotenoids.

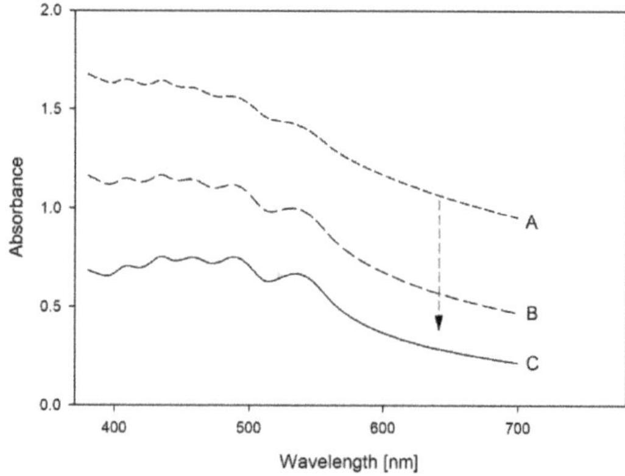

Figure 1. UV/Vis spectra of a carrot concentrate diluted in water (A), diluted in a water: sugar solution (1:1) (B), and diluted in a 69 °Bx sugar solution (C).

Carotenoid crystals display a pronounced absorbance peak at 520–545 nm while monomolecular carrot carotenoids exhibit an absorbance maximum at 440–460 nm (Figure 2) [9,10]. The change in absorbance at 440–460 nm in relation to the absorbance at 520–545 nm is thus further utilized in the qualitative determination of dissolution (increase of absorbance at 440–460 nm) as well as crystallization (increase of absorbance at 520–545 nm).

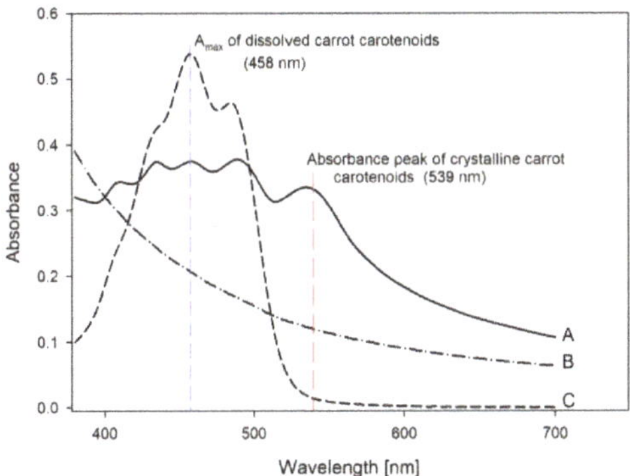

Figure 2. UV/Vis spectra of carrot carotenoids in sugar solution (A), the baseline measured after carotenoid extraction (B) and dissolved carrot carotenoids in sunflower oil (C).

For the analysis, 3 g of the samples were dissolved in 27 mL of distilled water and further diluted with a concentrated sugar solution by means of a laboratory blender (Waring 8011 EG). The final concentration of the sample in sugar solution (1.48 mg/mL) was constant for all samples analyzed in this study. The dilutions were transferred into centrifugal vials and centrifuged (rcf = 2800 g, 10 min) to eliminate air. An absorbance spectrum was recorded with a UV/Vis Spectrophotometer (Shimadzu 1800), from 380 to 700 nm, whereas the pure sugar solution was used as a blank. The UV/Vis spectra are a result of light absorption and scattering due to the pigments as well as the co-components in the sample. Figure 2 shows a typical UV/Vis spectrum of the carrot concentrate (Figure 2). To clearly identify the effect of the parameters studied on the physical state of the carotenoids, the absorption spectra were corrected for the absorbance and scattering through hydrophilic co-components (baseline in Figure 2). Qualitative information regarding the dissolution and crystallization of the carotenoids during processing and storage was derived from the comparison of the absorbance maxima at 440–460 nm and at 520–545 nm and from the shape of the UV/Vis absorbance curves.

2.7. Total Carotenoid Content

The total carotenoid content (TC) in the powders and the concentrate was determined spectrophotometrically after an extraction step in sunflower oil. Sample dilutions which had already been prepared (see previous section) in concentrated sugar solution, were consequently mixed with a known volume of sunflower oil (Cargill, Amsterdam, The Netherlands) in a laboratory blender for 90 s. To accelerate phase separation, the viscous mixture was transferred to centrifugal vials and centrifuged for 20 min (rcf = 2800 g, Eppendorf, Wesseling, Germany). After centrifugation the absorbance spectrum of the oil phase was recorded from 380 to 700 nm (Figure 2) and the absorbance at λ_{max} (458 nm) multiplied by the dilution was used to calculate the TC.

Initially, two methods were applied to quantify carotenoids in samples in the range of 0.2–2.2 mg/g powder. The first method was the extraction of the carotenoids from the sugar mixture and measurement in oil. The second method was carried out based upon the extraction procedure described by Sadler et al. [11], with some modifications. Briefly, 1–3 g of the sample was dissolved in 100 mL of distilled water prior to extraction. A total of 10 mL of extraction solvent (Acetone:Ethanol:Hexane, 1:1:2, 0,1% (*w/v*) BHT) were added to 2 mL of sample dilution and shaken for 1 min. After phase separation, the hexane phase was washed with 3 mL distilled water and transferred to a 25 mL volumetric flask. The extraction step was repeated at least two times by adding 5 mL of hexane. The hexane phases were combined in the volumetric flask and the volumetric flask was topped off with hexane to the mark and shaken before measuring the absorbance of the hexane phase in the spectrophotometer. Equation (1) was used to calculate the total carotenoid content.

$$TC\left[\frac{mg}{g}\right] = \frac{A \times V \times 10^3}{{}_{1\,cm}^{1\%}E \times 2\,mL \times m} \tag{1}$$

where A = absorbance at λ_{max}, V = the total volume of the extract (mL), ${}_{1\,cm}^{1\%}E$ = the average extinction coefficient of carotenoids in hexane 2500 [12] and m = the sample weight in g. The results from two methods correlated linearly (R^2 = 0.996) as shown in the Appendix A (Figure A1). Therefore, due to its non-toxic and non-volatile properties, oil was selected as a suitable extraction solvent. For calculations, the linear regression was used to calculate the TC in powders from absorbance values obtained from the carotenoid measurement in oil.

2.8. Surface Carotenoid (SC) and Encapsulated Carotenoids (EC)

The surface carotenoids (SC) were determined according to Wagner and Warthesen [13] with some adaptations. A total of 20–50 mg (m) of the powder sample were weighed into centrifugal vials and extracted with 10 mL hexane (0.1% *w/v* BHT). For the extraction of surface carotenoids, samples with the hexane were shaken for 5 min and subsequently centrifuged (rcf = 2800 g) for two min. The absorbance (A) of the hexane phase was measured spectrophotometrically at λ_{max} with hexane as sample blank. The SC were calculated according Equation (2).

$$SC\left[\frac{mg}{g}\right] = \frac{A \times 10^2}{{}_{1\,cm}^{1\%}E \times m}. \tag{2}$$

${}_{1\,cm}^{1\%}E$ and λ_{max} were equivalent to Equation (1). The amount of encapsulated carotenoids (EC) was defined as the difference of TC to SC.

2.9. Carotenoid Recovery, Encapsulation Efficiency, and Carotenoid Retention

To estimate the carotenoid losses during processing, the carotenoid recovery (CRec) was calculated according to Equation (3) from the theoretical carotenoid content in the liquid slurry formulation (TC_s) and the measured carotenoid content in the powder after drying (TC_p) in relation to the respective dry matter content of the powder (DM_p) and the slurry (DM_s).

$$CRec\,[\%] = \frac{TC_p}{DM_p} \times \left(\frac{TC_s}{DM_s}\right)^{-1} \times 100\%. \tag{3}$$

The encapsulation efficiency (EE) expresses the percentage of effectively encapsulated carotenoids after the drying process and can be calculated by Equation (4).

$$EE\,[\%] = \frac{EC}{TC_p} \times 100\%. \tag{4}$$

The carotenoid retention (CRet) after storage is the ratio of total carotenoids measured in the samples after 91 days of storage to the initial content which was measured 2–6 h after production.

2.10. Particle Size and Morphology

The particle size distribution (PSD) of the dried powders was determined by laser diffraction using a laser particles size analyzer with a powder feed unit (LA-960, Horiba, Kyoto, Japan). Particle morphology and carotenoid crystal distribution within the dried matrices were analyzed using a light microscope (Olympus BX51, Tokyo, Japan) with adapted camera system (Olympus XC50, Tokyo, Japan). The size of the carotenoid crystals was measured using the image analysis software cellSens Dimension (version 1.12) which enables size measurement during microscopy. Selected samples were examined in a benchtop scanning electron microscope (SEM) (JCM-6000, JEOL, Peobody, MA, USA) operating at 15 kV after sputter coating with gold for 30 s.

2.11. Storage Study

The produced powders were stored in closed PET containers (70 g/250 mL) with excess headspace and kept at 35 °C in a climate chamber (Ehret KBK 4200) for 91 days. The relative humidity in the climate chamber was 33% ± 4% during the storage period. The PET containers were opened and shaken weekly to avoid relevant oxygen decrease within the container. For oxygen-free storage, powders were filled into ceramic coated PET bags, flushed for 5–10 min with pure nitrogen, and sealed immediately. Oxygen sensors (OpTech®-O_2 Platinum, Mocon, Brooklyn Park, MN, USA) were placed in the packaging to monitor the oxygen concentration during storage.

2.12. Statistical Analysis

The experimental plan was based upon a modified design of experiments, including a variable number of independent process replicates, generally in duplicate and triplicate (Table 1). Sampling and analyses were carried out in triplicate. Statistical analyses were performed using STATGRAPHICS Centurion XVII, version 17.1.04 (Statpoint Technologies, Inc., Warrenton, VA, USA). Results of all parameters are expressed as mean ± standard deviation. A one way ANOVA ($p = 0.05$) and Tukey's HSD were used to determine statistical significant differences between processes and formulations. The standardized effect size of the tested measures on the carotenoid retention after storage (CRet) was estimated according to Glass' Δ whereas the reference powder (SD-Ref) was taken as control group [14].

3. Results and Discussion

3.1. Powder Particle Morphology and Component Distribution

Powders were analyzed for their morphology (SEM, light microscopy) and PSD to evaluate the impact of the drying technology on the powder structure (Figure 3). Particles produced by the spray drying process and collected after the cyclone separator were spherical, often exhibiting small dents and a mean diameter ($d_{4,3}$) ranging from 15 to 21 μm. The particle size and shape of the FD powder is determined by the milling process and resulted in angular particles with a mean diameter ($d_{4,3}$) of 112 ± 21 μm. The powder fraction, collected from the drying chamber of the spray dryer (SD-E), had a considerably larger particle size ($d_{4,3}$ 51–92 μm). The increased particle size can be explained by two factors: (a) the slower drying of bigger particles and therefore increased retention in the drying chamber and (b) particle agglomeration due to collision close to the drying wall [15,16]. SEM images confirmed the presence of both single spherical particles with a high particle diameter, and agglomerated fractions in SD-E powders.

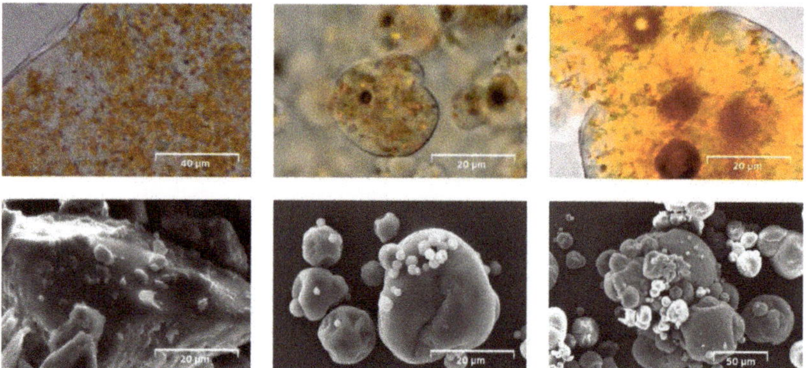

Figure 3. Microscopic images (top: light microcopy, bottom: SEM) of carrot concentrate powders: freeze-dried (FD; left), spray dried reference (SD-Ref; middle), and spray dried powder collected from the drying chamber (SD-E; right).

Light microscopic images revealed the presence of carotenoid crystals in the size range of 0.5–4 µm evenly dispersed in the carrier matrix. The detected amount of SC shows that considerable fractions of the carotenoids, were not effectively encapsulated in the matrix, but present on the particle surface after drying (Tables 2 and 3).

Table 2. Carotenoid recovery (CRec), surface carotenoid (SC) content after production, and carotenoid retention (CRet) after 91 days storage (35 °C, air) of carrot concentrate powders produced without added antioxidants.

Sample Code	CRec (%)	SC (%)	CRet (%)
SD-Ref	99.4 ± 0.8 [e]	29.2 ± 1.3 [e,d]	44.9 ± 0.8 [c]
FD	98.9 ± 1.3 [c,d,e]	26.3 ± 3.9 [d,e]	12.3 ± 2.1 [a]
SD-no-HPH	98.3 ± 0.4 [c,d]	32.9 ± 0.8 [f]	38.1 ± 0.5 [b]
SD-E	97.1 ± 0.5 [a,b]	14.1 ± 2.5 [a]	66.0 ± 5.4 [h]
SD 50% GA	98.8 ± 0.7 [d,e]	23.6 ± 1.0 [c,d]	52.6 ± 1.2 [d]
SD 100% GA	95.9 ± 0.8 [a]	19.6 ± 0.6 [b,c]	60.1 ± 1.3 [e]
SD 50% OSA	99.4 ± 0.5 [e]	16.3 ± 0.6 [a,b]	54.6 ± 1.2 [d]
SD 100% OSA	98.0 ± 0.5 [b,c,d]	12.2 ± 0.1 [a]	61.6 ± 1.3 [e]

Means within the same column followed by different letters are statistically significant different ($p \leq 0.05$).

Table 3. Carotenoid recovery (CRec), surface carotenoid (SC) content after production, and carotenoid retention (CRet) after 91 days storage (35 °C, air) of carrot concentrate powders produced with different levels of antioxidants.

Sample Code	CRec (%)	SC (%)	CRet (%)
SD-Ref [1]	99.4 ± 0.8 [c]	29.2 ± 1.3 [c,d]	44.9 ± 0.8 [a]
SD-Toc-low	97.8 ± 0.2 [a,b]	25.4 ± 0.6 [a]	52.5 ± 0.5 [c]
SD-SA-low	96.8 ± 0.5 [a,b]	26.5 ± 0.8 [a,b]	49.7 ± 0.6 [b]
SD-Toc-high	97.2 ± 1.0 [a]	28.5 ± 1.0 [b,c]	59.8 ± 1.8 [e]
SD-SA-high	97.7 ± 0.5 [a,b]	25.1 ± 0.7 [c]	54.1 ± 1.9 [c,d]
SD-Toc-SA-low	98.8 ± 1.2 [b,c]	28.2 ± 2.0 [b,c]	49.3 ± 0.4 [b]
SD-Toc-SA-high	97.8 ± 1.4 [a,b]	30.1 ± 0.3 [d]	60.2 ± 1.3 [e]

[1] SD-Ref values are included for direct comparison and represent the same dataset as in Table 2. Means within the same column followed by different letters are statistically significant different ($p \leq 0.05$).

3.2. Impact on Processing on Carrot Carotenoid Content and UV/Vis Absorbance

The recovery of total carotenoids (CRec) in the produced powders can be derived from Tables 2 and 3. A high carotenoid recovery of >95% was measured in FD and SD powders, indicating a high stability of the carrot carotenoids throughout both drying processes. CRec was significantly lower ($p \leq 0.05$) in the powder collected from the spray-drying chamber (SD-E), although the small difference compared to SD-Ref was surprising (<3%) considering that SD-E was continuously exposed to the hot air flow during powder production.

These results are in contrast with high degradation rates of carotenoids during sample production reported by some authors. In spray drying β-carotene crystal dispersions with maltodextrin as a carrier, Desobry et al. [17] measured a process loss of total carotenoids of 11%. Even higher values of up to 85% carotenoid degradation were ascribed to the spray drying process when β-carotene nano-emulsion was dried, in a recent study [18]. Nevertheless, the high CRec is reasonable as carotenoids degrade typically upon high or prolonged heat impact, which is limited during spray and freeze drying when product recovery is rapid after drying. Additionally, the presence of carrot derived antioxidants such as tocopherols as well as the initial supramolecular structure, might increase the stability of carrot carotenoids during processing [7,19].

UV/Vis spectra derived from measurement of the carrot concentrates in concentrated sugar solution, corresponded well with the in situ UV/Vis spectra of carotenoids in carrot tissue [8]. The difference between the spectra of the concentrate and the powders were generally low, indicating that the carotenoids retained their naturally occurring state during processing. Variations were most pronounced between powders produced by the two different drying technologies. In Figure 4 the differences in the UV-Vis spectra of a FD powder and a SD-Ref is shown. At equal carotenoid concentrations (TC = 1.94 ± 0.02 mg/g and 1.93 ± 0.3 mg/g in the SD and FD powder respectively), absorbance intensity of the monomeric carotenoid fraction SD powders compared to FD powders, while FD powders showed a higher absorbance around 539 nm. The observed changes from FD to SD powders suggest an increased dissolution of β-carotene crystals in the SD samples [7].

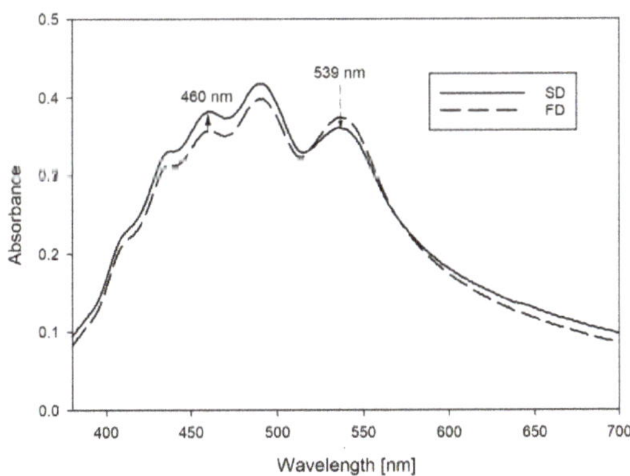

Figure 4. UV/Vis spectra of carrot concentrate powder after spray drying (solid line) and freeze drying (dashed line).

3.3. Impact of Ambient Oxygen on the Carotenoid Degradation during Storage

To determine the impact of ambient oxygen on carotenoid degradation during storage, samples with and without antioxidants were additionally stored in a nitrogen atmosphere. The oxygen level in the nitrogen flushed packaging was below 1% for all discussed results. The CRet in SD-Ref was 97.0% ± 0.5% after 91 days storage at 35 °C, which was surprisingly high considering that nitrogen flushing was not expected to remove residual oxygen within the particle vacuoles and pores. No decrease in carotenoid content was detected in powders stored in the nitrogen flushed packaging containing low levels of mixed AO (SD-Toc-SA-low). In contrast, CRet was below 66.0% ± 5.4% for all powders stored at ambient atmosphere as shown in Tables 2 and 3. The availability of external oxygen can thus be regarded as main factor in the initiation and promotion of carotenoid degradation during storage at 35 °C. This is in agreement with observations made by Stevanovich and Karel [20] who investigated β-carotene degradation kinetics in dry model systems. The authors concluded in their study that oxygen diffusion through the different layers of the dry particle is a limiting factor for carotenoid degradation.

Despite the high stability at oxygen free storage, the presence of lipid dissolved oxygen, and oxygen enclosed within particle pores cannot be excluded in the tested carrot concentrate powders. Both factors have shown to accelerate lipid oxidation during the storage in SD powders [21]. However, the effect on the CRet was negligible in the tested carrot concentrate powders. Endogenous antioxidants of the carrot components [19] might have protected the carotenoids within the powders from the effect of dissolved or enclosed oxygen or other reactive species and thus contributed to the measured CRet.

While no significant changes in the TC were detected in SD-Toc-SA-low samples, when stored with exclusion of ambient oxygen, small changes in the macromolecular conformation could be derived from the comparison of UV-Vis spectra, before and after the storage at 35 °C (Figure 5). A simultaneous increased absorbance after storage at 539 nm and decrease in absorbance at 450–460 nm was observed. The observed shift was low, but significant ($p \leq 0.05$) and is a strong indicator for aggregation of carotenoid molecules [22]. We therefore conclude that some carotenoids which are monomolecular after spray drying, aggregate or assemble to crystals during storage in the carrot concentrate powders. However, the observed effect was minor and implications for the bioavailability and color hue need to be assessed in order to estimate the relevance of the carotenoid aggregation for product quality.

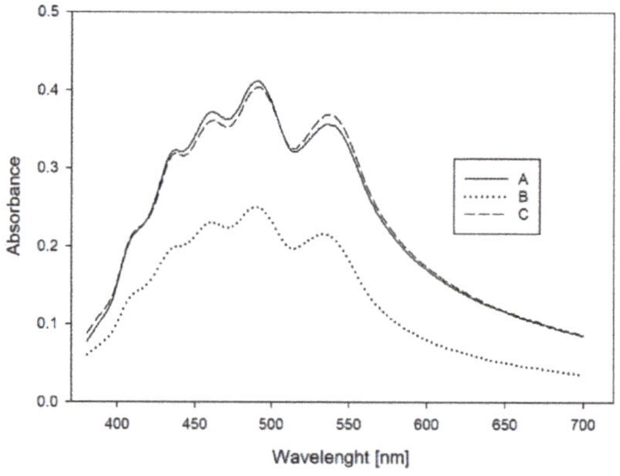

Figure 5. UV/Vis spectra of a spray dried carrot concentrate powder containing mixed tocopherols and sodium ascorbate as antioxidants (SD-Toc-SA-low) after production (A), after storage in air (B) and after storage in nitrogen flushed packaging (C) for 91 days at 35 °C.

3.4. Encapsulation Efficiency and Drying Process as Impact Factors for Carotenoid Retention during Storage

When stored under ambient atmosphere, degradation of surface carotenoids (SC) as well as encapsulated carotenoids occurred in all samples, with a significant faster degradation of SC ($p \leq 0.05$). In samples produced with MD as carrier and without additional AO, the amount of remaining SC after storage ranged from 2.6% to 4.9% of the initial value. Although slower, also carotenoids enclosed in the particle matrix (EC) degraded. This implies that oxygen diffusion must also have occurred throughout the particle matrix, initiating degradative reactions of EC in all tested samples. Figure 6 shows the carotenoid retention of powders produced without antioxidants as a function of encapsulation efficiency. CRet increased with increasing EE of spray dried particles but varied widely between particles produced by differing drying methods.

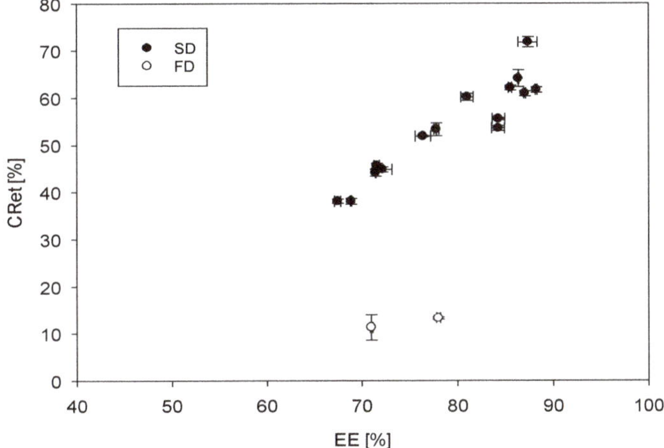

Figure 6. Carotenoid retention (CRet) in carrot concentrate powders after storage (91 days, 35 °C, air) as a function of the encapsulation efficiency (EE). Each data point represents the mean of a triplicate analysis. Error bars fall below the size of the symbols in some cases.

After storage (91 days, 35 °C, air), 87.5% ± 1.3% of the carotenoids were degraded in FD powders while only 44.9% ± 0.8% were degraded in the SD-Ref. Other authors have shown that the microstructure and inner porosity of FD particles, influences the diffusion of oxygen through the dried particle and thus the oxidation of an encapsulated component [23]. The fast oxidation of carotenoids in the FD matrix compared to the SD powder is likely to be caused by increased mobility of external oxygen in the porous FD particle [24]. Contrary, carotenoids in powders collected from the drying chamber of the spray dryer (SD-E) showed a significant higher stability during storage compared to the powders collected after the cyclone separator ($p \leq 0.05$). This result was surprising, since SD-E powders were subjected to increased heat during production. However, SD-E powder particles were also significantly larger compared to powder particle collected after the cyclone separator. The superior size of SD-E particles, most likely, provided increased protection to oxygen diffusion during storage, due to their lower specific surface area, and increased particle wall diameter [25]. Additionally, Maillard reaction products might have been produced during the prolonged heat impact, which would exhibit antioxidant activity, resulting in a combined effect of newly formed antioxidants and improved physical barrier in SD-E powders [26].

3.5. Impact of Functional Additives on the Stability of Carotenoids in Carrot Concentrate Powders

3.5.1. Effect of Surface-Active Carrier

The substitution of 50% and 100% of MD with a surface-active carrier (GA or OSA-starch), generally improved the EE, as well as the CRet (Table 2). GA and OSA-starch have shown to positively influence the physical properties carotenoid-rich spray dried plant extracts and concentrates [27]. However, the reported effect on EE and shelf life of plant concentrate powders in comparison to a non-surface active maltodextrin varies widely and is thus currently not conclusive [28].

The substitution of MD with OSA-starch and GA improved the EE and CRet significantly with higher concentration (100% substitution) being more beneficial. The fraction of surface carotenoids was reduced from 29.2% ± 1.3% (*w/w*) for maltodextrin-based powders to 19.6% ± 0.6% (*w/w*) and 12.2% ± 0.1% (*w/w*) GA and OSA-starch respectively, which further influenced the retention of total carotenoids (CRet) during storage (Table 2). OSA-starch was more effective in reducing surface carotenoids compared to GA, whereby the higher resulting EE did not lead to a higher CRet after 91 days storage.

In the encapsulation of active components by means of emulsification and spray drying, GA and OSA-starch have shown to improve the oxidation stability compared to maltodextrin, which is ascribed to their function as emulsifiers, their film forming properties during the drying process and in cases superior barrier properties [29,30]. Both, OSA-starch and GA contain a polysaccharide backbone and lipophilic side chains which attach to the lipid/water interface and enable the production of emulsions with small droplet size.

A small droplet size distribution of emulsions prior to spray drying has shown to be a main impact factor for EE with smaller droplets leading to a higher EE [31]. In SD emulsions which are stabilized by additional emulsifiers, GA has shown to be of no advantage concerning EE and oxidation stability compared to MD [32]. This indicates that the emulsifying properties of the surface active material are mainly beneficial in systems lacking other surface active components. A corresponding mechanism of GA and OSA-starch for the stabilization of crystalloid or particulate active ingredients during spray drying is likely but cannot be derived from current scientific literature. The proposed stabilizing mechanism of OSA-starch and GA in the carrot concentrate powders is the structural stabilization of small carotenoid crystals or residue lipid fractions during the processing. While this hypothesis is supported by the lower amount of SC in samples with GA and OSA-starch, a smaller PSD of the carotenoid containing phase in the respective samples was not unambiguously established.

3.5.2. Effect of Antioxidant Addition

The CRet of the SD sample produced with various levels of antioxidants are shown in Table 3. Antioxidants reduced the carotenoid degradation in encapsulated carotenoids (EC) and surface carotenoid (SC) fraction, resulting in significantly higher CRet after storage in samples produced with AO ($p \leq 0.05$). Notably, the effect was more pronounced in SC compared to EC. The addition of antioxidants decreased the degradation of SC resulting in a retention of SC ranging from 8.1% to 44.5% compared to 2%–4% in samples without AO. The highest retention of SC was measured in SD-Toc-SA-high samples (44.5% ± 1.2%) and the lowest in SD-SA-low (8.1% ± 2.0%), indicating that the high levels chosen for this study were effective for reducing the oxidation of the most exposed SC. Surprisingly, the retention of EC in samples produced with low levels of antioxidants did not differ significantly ($p \leq 0.05$) from the retention of EC in the SD-Ref and only an increase from 62.8% ± 2.5% in SD-Ref to a maximum of 74.2% ± 3.2% in samples produced with high levels of antioxidants was observed. Similar observations were made by Velasco et al. [33] who tested the effectivity of an antioxidant system (ascorbic acid, lecithin, and tocopherol) on encapsulated and surface fat of spray dried fish oil emulsions. In their study, the tested antioxidant system effectively reduced lipid oxidation of the surface oil fraction while no significant effect on the encapsulated oil droplets could be observed. Further studies of this working group support their initial observation of the heterogeneous aspect of lipid oxidation in dried dispersed systems [34].

An additive or synergistic effect of Toc and SA, which is often described for bulk oil or emulsion systems, was not observed in this study. After the storage period, the UV/Vis spectra of samples with high amounts of antioxidants showed significant differences in the absorbance at 460 nm at similar carotenoid concentrations, compared to samples spray dried with a surface-active carrier when measured in sugar solution. In Figure 7 the UV/Vis spectra of a sample produced with Toc and SA and a sample produced with OSA-starch as carrier material are compared. The sample containing antioxidants shows a pronounced higher absorbance at 460 nm compared to the sample stabilized with OSA-starch and GA, which cannot be explained by a higher total carotenoid content. A possible explanation would be that the tested antioxidants have a stronger protective effect on the carotenoid fraction dissolved in the carrot concentrate powder compared to the crystalline fraction.

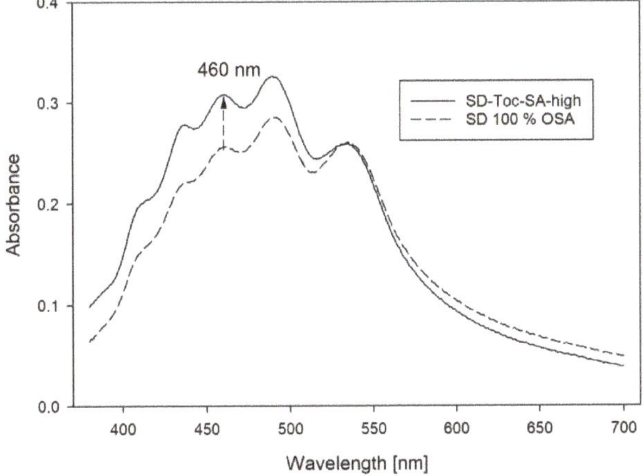

Figure 7. Variation in UV/Vis spectra of spray dried carrot carotenoids formulated with antioxidants (mixed tocopherols and sodium ascorbate) and maltodextrin as carrier material (SD-Toc-SA-high) and no antioxidants and octenyl succinic anhydride-modified starch as carrier material (SD 100% OSA) after storage (35 °C, 91 days, air).

The mechanism and synergism of the tested antioxidant system was thoroughly investigated for bulk oils and emulsions while comprehensive data on their efficacy and mechanism of action in dried systems is still deficient [35]. Recent studies on the mechanism of anti- and pro-oxidants in dried dispersed systems indicate that their respective mechanism and efficacy can vary widely compared to liquid systems [36,37]. Possible reasons are the decreased mobility for all components, the increased role of physical barriers which govern oxygen diffusion, and the increased contact area with ambient oxygen in dry products. In a complex food system like dried carrot concentrate, variable factors can impact the lipid or carotenoid oxidation including interfacial areas between carotenoid crystals and the carrier and an uneven distribution of antioxidants throughout the system.

4. Conclusions

Crystalline carrot carotenoids showed high stability during slurry processing and drying, followed by pronounced degradation during storage in the presence of oxygen. Of the tested strategies for carotenoid stabilization, the exclusion of oxygen clearly had the most profound effect on carotenoid stability during storage. In conclusion, carotenoid degradation in differently stored samples was mainly promoted by ambient oxygen and its diffusion through the matrix. Freeze drying had the most pronounced, but negative, effect on the storage stability of the powders when it was used as alternative technology instead of spray drying. A possible explanation is the increased inner porosity of FD samples compared to SD samples which provides lower protection for the carotenoids towards ambient oxygen. The large differences in carotenoid retention in SD and FD powders indicate a high impact of the particle morphology on the embedded carotenoids.

Added functional ingredients were effective in reducing the degradation of crystalline carotenoids but showed clear limitations. The substitution of MD as standard carrier with a surface-active carrier (GA or OSA-starch) reduced the initial amount of surface carotenoids (SC) which resulted a higher carotenoid retention during storage. However, even in samples with solely GA or OSA-starch more than 38% of the carotenoids were degraded after 3-month storage (35 °C) demonstrating that additional protective measures are necessary for encapsulated carotenoids in order to inhibit degradation. Antioxidants showed to be effective to improve CRet, especially when applied in high concentration of 2500–5000 µg/g, but were not able to inhibit carotenoid degradation when exposed to ambient oxygen. Further research is needed in order to elucidate the interaction of carotenoid crystals in a dried, dispersed system and added antioxidants as well as to find optimized concentrations. In order to reach a higher shelf-stability at ambient environment of functional carrot concentrate powders, a combined approach with an optimized particle morphology as well as carrier and antioxidant system should be considered.

This study successfully revealed the role of different impact factors for the stabilization of carotenoids in powders made from vegetable concentrates. Furthermore, an appropriate method for the analysis and identification of crystalline carotenoids within the complex matrix of plant concentrates was developed. The results obtained form the basis for further research and formulation for the production of high-quality functional food systems.

Author Contributions: Conceptualization, K.H., M.V., A.K. and H.J., methodology, K.H. and A.K.; sample production and formal analysis, K.H. and P.R.; data curation, K.H. and H.J.; writing—original draft preparation, K.H.; writing—review and editing, K.H., A.K., H.J., M.V.; supervision, M.V.; project administration, H.J.

Funding: Part of the equipment used in this study was financed by EQ-BOKU VIBT GmbH and belongs to the Center for Preservation and Aseptic Processing.

Conflicts of Interest: A.K. and M.V. were employed at GNT-Europa GmbH during their engagement in this study. The company GNT-Europe GmbH is also a producer of the coloring carrot concentrates which were used in this study. Both co-authors contributed to the conceptualization of the study. Their engagement played no role in the analysis, interpretation, and selection of the data. P.R. was employed at GNT International GmbH after he contributed at the University of Natural Resources and Life Sciences (BOKU).

Appendix A

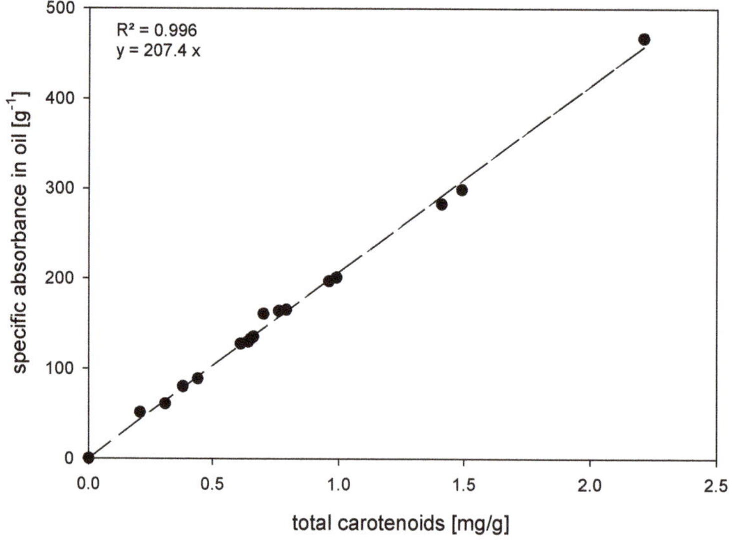

Figure A1. Correlation between the specific absorbance of carrot concentrate powders after extraction in sunflower oil and total carotenoids determined by solvent extraction in hexane. Sample points represent the mean of a triplicate analysis.

References

1. Boon, C.S.; McClements, D.J.; Weiss, J.; Decker, E. Factors influencing the chemical stability of carotenoids in foods. *Crit. Rev. Food Sci. Nutr.* **2010**, *50*, 515–532. [CrossRef] [PubMed]
2. McClements, D.J. Crystals and crystallization in oil-in-water emulsions: Implications for emulsion-based delivery systems. *Adv. Colloid. Interface Sci.* **2012**, *174*, 1–30. [CrossRef] [PubMed]

3.   Köhn, S.; Kolbe, H.; Korger, M.; Köpsel, C.; Mayer, B.; Auweter, H.; Lüddecke, E.; Bettermann, H.; Martin, H. Aggregation and Interface Behaviour of Carotenoids. In *Carotenoids Volume 4: Natural Functions*; Britton, G., Liaaen-Jensen, S., Pfander, H., Eds.; Birkhäuser Basel: Basel, Switzerland, 2008; pp. 53–98.

4.   Schweiggert, R.; Mezger, D.; Schimpf, F.; Steingass, C.B.; Carle, R. Influence of chromoplast morphology on carotenoid bioaccessibility of carrot, mango, papaya, and tomato. *Food Chem.* **2012**, *135*, 2736–2742. [CrossRef] [PubMed]

5.   Kläui, H.; Bauernfeind, J.C. Carotenoids as food colors. In *Carotenoids as Colorants and Vitamin a Precursors*; Bauernfeind, C.J., Ed.; Academic Press: San Diego, CA, USA, 1981; pp. 47–317.

6.   Knockaert, G.; Lemmens, L.; Buggenhout, S.V.; Hendrickx, M.; Loey, A. Changes in β-carotene bioaccessibility and concentration during processing of carrot puree. *Food Chem.* **2012**, *133*, 60–67. [CrossRef]

7.   Marx, M.; Stuparic, M.; Schieber, A.; Carle, R. Effects of thermal processing on trans–cis-isomerization of β-carotene in carrot juices and carotene-containing preparations. *Food Chem.* **2003**, *83*, 609–617. [CrossRef]

8.   Hempel, J.; Müller-Maatsch, J.; Carle, R.; Schweiggert, R.M. Non-destructive approach for the characterization of the in situ carotenoid deposition in gac fruit aril. *J. Food Compos. Anal.* **2018**, *65*, 16–22. [CrossRef]

9.   Gaier, K.; Angerhofer, A.; Wolf, H.C. The lowest excited electronic singlet states of all-trans β-carotene single crystals. *Chem. Phys. Lett.* **1991**, *187*, 103–109. [CrossRef]

10. Simonyi, M.; Bikadi, Z.; Zsila, F.; Deli, J. Supramolecular Exciton Chirality of Carotenoid Aggregates. *Chem. Inform.* **2004**, *35*. [CrossRef]

11. Sadler, G.; Davis, J.; Dezman, D. Rapid Extraction of Lycopene and β-Carotene from Reconstituted Tomato Paste and Pink Grapefruit Homogenates. *J. Food Sci.* **1990**, *55*, 1460–1461. [CrossRef]

12. Rodriguez-Amaya, D.B. Quantitative analysis, in vitro assessment of bioavailability and antioxidant activity of food carotenoids—A review. *J. Food Compos. Anal.* **2010**, *23*, 726–740. [CrossRef]

13. Wagner, L.A.; Warthesen, J. Stability of Spray-Dried Encapsulated Carrot Carotenes. *J. Food Sci.* **1995**, *60*, 1048–1053. [CrossRef]

14. Hedges, L.V. Distribution Theory for Glass's Estimator of Effect size and Related Estimators. *J. Educ. Stat.* **1981**, *6*, 107–128. [CrossRef]

15. Gianfrancesco, A.; Turchiuli, C.; Dumoulin, E. Powder agglomeration during the spray-drying process: Measurements of air properties. *Dairy Sci. Technol.* **2008**, *88*, 53–64. [CrossRef]

16. Vicente, J.; Pinto, J.; Menezes, J.; Gaspar, F. Fundamental analysis of particle formation in spray drying. *Powder Technol.* **2013**, *247*, 1–7. [CrossRef]

17. Desobry, S.A.; Netto, F.M.; Labuza, T.P. Preservation of beta-carotene from carrots. *Crit. Rev. Food Sci. Nutr.* **1998**, *38*, 381–396. [CrossRef] [PubMed]

18. Chen, J.; Li, F.; Li, Z.; McClements, D.J.; Xiao, H. Encapsulation of carotenoids in emulsion-based delivery systems: Enhancement of β-carotene water dispersibility and chemical stability. *Food Hydrocoll.* **2017**, *69*, 49–55. [CrossRef]

19. Sharma, K.D.; Karki, S.; Thakur, N.S.; Attri, S. Chemical composition, functional properties and processing of carrot—A review. *J. Food Sci. Technol.* **2012**, *49*, 22–32. [CrossRef] [PubMed]

20. Stefanovich, A.F.; Karel, M. Kinetic of beta-carotene degradation at temperatures typical of air drying of foods. *J. Food Process Pres.* **1982**, *6*, 227–242. [CrossRef]

21. Serfert, Y.; Drusch, S.; Schmidt-Hansberg, B.; Kind, M.; Schwarz, K. Process engineering parameters and type of n-octenylsuccinate-derivatised starch affect oxidative stability of microencapsulated long chain polyunsaturated fatty acids'. *J. Food Eng.* **2009**, *95*, 386–392. [CrossRef]

22. Adamkiewicz, P.; Sujak, A.; Gruszecki, W.I. Spectroscopic study on formation of aggregated structures by carotenoids: Role of water. *J. Mol. Struct.* **2013**, *1046*, 44–51. [CrossRef]

23. Harnkarnsujarit, N.; Charoenrein, S.; Roos, Y.; Porosity, H. Water Activity Effects on Stability of Crystalline β-Carotene in Freeze-Dried Solids. *J. Food Sci.* **2012**, *77*, E313–E320. [CrossRef] [PubMed]

24. Anwar, S.H.; Kunz, B. The influence of drying methods on the stabilization of fish oil microcapsules: Comparison of spray granulation, spray drying, and freeze drying. *J. Food Eng.* **2011**, *105*, 367–378. [CrossRef]

25. Soottitantawat, A.; Bigeard, F.; Yoshii, H.; Furuta, T.; Ohkawara, M.; Linko, P. Influence of emulsion and powder size on the stability of encapsulated d-limonene by spray drying. *Innov. Food Sci. Emerg. Technol.* **2005**, *6*, 107–114. [CrossRef]

26. Vhangani, L.; Van, N.; Wyk, J. Antioxidant activity of Maillard reaction products (MRPs) derived from fructose–lysine and ribose–lysine model systems. *Food Chem.* **2013**, *137*, 92–98. [CrossRef] [PubMed]
27. Janiszewska-Turak, E. Carotenoids microencapsulation by spray drying method and supercritical micronization. *Food Res. Int.* **2017**, *99*, 891–901. [CrossRef] [PubMed]
28. Soukoulis, C.; Bohn, T. A Comprehensive Overview on the Micro- and Nano-technological Encapsulation Advances for Enhancing the Chemical Stability and Bioavailability of Carotenoids. *Crit. Rev. Food Sci. Nutr.* **2015**. [CrossRef] [PubMed]
29. Jafari, S.M.; Assadpoor, E.; He, Y.; Bhandari, B. Encapsulation Efficiency of Food Flavours and Oils during Spray Drying. *Dry. Technol.* **2008**, *26*, 816–835. [CrossRef]
30. Krishnan, S.; Bhosale, R.; Singhal, R. Microencapsulation of cardamom oleoresin: Evaluation of blends of gum arabic, maltodextrin and a modified starch as wall materials. *Carbohydr. Polym.* **2005**, *61*, 95–102. [CrossRef]
31. Munoz-Ibanez, M.; Nuzzo, M.; Turchiuli, C.; Bergenståhl, B.; Dumoulin, E.; Millqvist-Fureby, A. The microstructure and component distribution in spray-dried emulsion particles. *Food Struct.* **2016**, *8*, 16–24. [CrossRef]
32. Kang, Y.; Lee, Y.; Kim, Y.K.; Chang, Y.H. Characterization and storage stability of chlorophylls microencapsulated in different combination of gum Arabic and maltodextrin. *Food Chem.* **2019**, *272*, 337–346. [CrossRef]
33. Velasco, J.; Dobarganes, M.C.; Márquez-Ruiz, G. Oxidation of free and encapsulated oil fractions in dried microencapsulated fish oils. *Grasas Aceites* **2000**, *51*, 6. [CrossRef]
34. Velasco, J.; Marmesat, S.; Dobarganes, C.; Márquez-Ruiz, G. Heterogeneous Aspects of Lipid Oxidation in Dried Microencapsulated Oils. *J. Agric. Food Chem.* **2006**, *54*, 1722–1729. [CrossRef] [PubMed]
35. Berdahl, D.R.; Nahas, R.I.; Barren, J.P. Synthetic and natural antioxidant additives in food stabilization: Current applications and future research. In *Oxidation in Foods and Beverages and Antioxidant Applications*; MCClements, D.J., Decker, E.A., Eds.; Woodhead Publishing: Sawston, England, 2010; pp. 272–320.
36. Baik, M.Y.; Suhendro, E.L.; Nawar, W.W.; McClements, D.J.; Decker, E.A.; Chinachoti, P. Effects of antioxidants and humidity on the oxidative stability of microencapsulated fish oil. *J. Am. Oil Chem. Soc.* **2004**, *81*, 355–360. [CrossRef]
37. Barden, L.; Decker, E.A. Lipid Oxidation in Low-moisture Food: A Review. *Crit. Rev. Food Sci.* **2016**, *56*, 2467–2482. [CrossRef] [PubMed]

Article

Influence of Pulsed Electric Field and Ohmic Heating Pretreatments on Enzyme and Antioxidant Activity of Fruit and Vegetable Juices

Cinzia Mannozzi [1,2,*], Kamon Rompoonpol [1], Thomas Fauster [1], Urszula Tylewicz [2], Santina Romani [2], Marco Dalla Rosa [2] and Henry Jaeger [1]

[1] Institute of Food Technology, University of Natural Resources and Life Sciences (BOKU), Muthgasse 18, 1190 Vienna, Austria
[2] Department of Agricultural and Food Science, University of Bologna, Piazza Goidanich 60, 47521 Cesena, Italy
* Correspondence: cinzia.mannozzi2@unibo.it

Received: 29 May 2019; Accepted: 6 July 2019; Published: 8 July 2019

Abstract: The objective of this work was to optimize pulsed electric field (PEF) or ohmic heating (OH) application for carrot and apple mashes treatment at different preheating temperatures (40, 60 or 80 °C). The effect of tissue disintegration on the properties of recovered juices was quantified, taking into account the colour change, the antioxidant activity and the enzyme activity of peroxidase (POD) in both carrot and apple juice and polyphenol oxidase (PPO) in apple juice. Lower ΔE and an increase of the antioxidant activity were obtained for juice samples treated with temperature at 80 °C with or without PEF and OH pretreatment compared with those of untreated samples. The inactivation by 90% for POD and PPO was achieved when a temperature of 80 °C was applied for both carrot and apple mash. A better retention of plant secondary metabolites from carrot and apple mashes could be achieved by additional PEF or OH application. Obtained results are the basis for the development of targeted processing concepts considering the release, inactivation and retention of ingredients.

Keywords: PEF; OH; POD; colour; extraction

1. Introduction

Low-intensity pulsed electric field (PEF) can enhance the mass transfer during extraction by increasing cell membrane permeability, known as electroporation. Therefore, PEF treatments can enhance the release of specific intracellular compounds from plant tissues [1–4].

At the same time, ohmic heating (OH) as an alternative thermal pretreatment could be used prior to extraction. The volumetric energy dissipation and the rapid and uniform heating represent advantages, especially for viscous particulate products such as fruit or vegetable mash [5]. In addition, the short processing times during the OH treatment may cause less degradation of colour- and heat-sensitive substances.

In plant cells, antioxidant compounds are mainly located in the vacuoles, whereas the enzymes peroxidase (POD) and polyphenoloxidase (PPO) are found in plastids [6].

Processing of plant tissues compromises the internal compartmentalization that allows the contact between degradative enzymes and their substrates (phenolic compounds), implying the reaction known as enzymatic browning. In the case of POD, phenolic compounds are oxidized at the expense of H_2O_2, leading to flavour changes [7]. Instead, PPO is an oxidoreductase which catalyses the oxidation of phenolic compounds in o-quinones, which are subsequently polymerized into brown pigments [8]. Therefore, the inactivation of POD and PPO enzymes is a crucial prerequisite and indicator of quality in the processing of fruit and vegetables.

Moreover, the activation of enzymes, including an increased release and the enhancement of enzymatic reactions by the cell disintegration applied at early stages during fruit and vegetable processing might be a limitation for the shelf-life of recovered juices. Thermal treatment has been used in order to reduce the enzyme activity, but it causes negative effects on quality and related nutritional compounds [9]. Nonthermal food preservation technologies are considered to be more efficient in terms of required energy and in terms of avoiding heat-induced changes of colour, flavour and nutritional value [10]. However, enzyme inactivation achieved during the nonthermal preservation of juices is rather limited [11].

Carrot and apple are good sources of carotenoids and phenols, which are located in the chromoplasts and in the vacuoles of the plant cells, respectively [12,13]. They are considered to provide health benefits due to the antioxidant properties that also contribute to the colour and sensory quality of fresh and processed products [14].

To promote the selectivity of the extraction of bioactive compounds from plant tissues, pulsed electric field [15–17] and ohmic heating [18,19] treatment have been already investigated.

Bhat et al. [18] applied thermal and OH treatment (60–90 °C; 1–5 min) to bottle gourd and compared the effects on total phenolic content and colour of obtained juices. The total phenolic content increased with OH and thermal application at 80 °C for 4 min and 90 °C for 5 min, respectively, and the best colour retention was observed for OH-treated juice at 80 and 90 °C.

Saxena et al. [20] reported the effect of OH treatment on PPO activity in sugarcane juice. A high PPO inactivation was observed by applying 32 V/cm at 90 °C for 5 min.

However, for the optimization of the PEF and OH process parameters with regard to reducing the energy requirement and process time and increasing yield and quality, more information is required. Subsequently, the impact on enzyme activity and the recovery of bioactive compounds from the raw material need to be investigated. A first part of the study focused on juice yield and selected ingredients extraction by taking into account the induced cell disintegration effects [21]. This second part of the work aimed at understanding the effects of PEF and OH treatments on antioxidant properties, colour and enzyme activity such as peroxidase (POD) for both juices and polyphenoloxidase (PPO) for apple juice. The optimization of the two processing technologies was performed by the modulation of process parameters as well as treatment temperatures, by applying a preheating step (40, 60 or 80 °C) in order to evaluate the effects of thermal and electric field on antioxidant and enzyme activity of recovered apple and carrot juices.

2. Materials and Methods

2.1. Plant Raw Material and Mash Preparation

Raw materials (carrots and apples) were bought from the local market in Vienna (Austria). They were washed and cut into smaller pieces. A mill (Alexanderwerk, Remscheid, Germany) with replaceable stainless-steel screens with a grinding level of 2 mm for carrots and 5 mm for apples was used in order to produce the mash.

2.2. Mash Pretreatment and Juice Production

For PEF treatment of apple and carrot mash, a batch PEF system (DIL, Quakenbrück, Germany) was used. The distance between parallel plate electrodes in the treatment chamber was set to 5 cm. 50 exponential decay pulses (discharge capacity 0.5 µF, pulse energy 4 J, pulse width 10 µs) were applied to 400 g of mash. The output voltage was set to 4 kV in order to achieve in the treatment chamber an electric field strength of 0.8 kV/cm. A total specific energy input (Wspecific) of 0.5 kJ/kg was applied. The total treatment time of 0.5 ms was calculated by multiplying the pulse width with the number of pulses.

OH treatment was performed in the same treatment chamber as for the PEF treatment by using a generator (DIL, Germany) to apply the electric field (1.1 A, 572 V, 12 kHz, 0.6 kW), resulting in an electric field of 114 V/cm. Different temperature–time profiles were acquired depending on the selected temperatures for the different carrot and apple mashes [21].

For preheating to the different initial temperatures (40, 60 and 80 °C), microwave (MT 267, Whirlpool, München, Germany) heating with a power of 850 W was applied for different predefined times. Temperatures were measured with a PT100 thermocouple directly during the treatment [21].

After the different pretreatment, all mash batches were cooled to room temperature and pressed using a manual laboratory juice press (Hafico, Germany) with textile cloth; eleven juice samples were obtained for both carrot and apple, in three replicates each (Table 1).

Table 1. Overview on mash-treatment conditions applied for apple and carrot mash. Pulsed electric field (PEF), ohmic heating (OH).

Treatment	Sample	Wspecific (kJ/kg)
Untreated	Control	0
PEF at 20 °C	PEF (20 °C)	0.5
Preheating 40 °C + PEF	40 °C-PEF	192.5
Preheating 60 °C + PEF	60 °C-PEF	382.5
Preheating 80 °C + PEF	80 °C-PEF	765.5
OH from 20 °C to 40 °C	(20 °C–40 °C) OH	110
OH from 20 °C to 60 °C	(20 °C–60 °C) OH	222
OH from 20 °C to 80 °C	(20 °C–80 °C) OH	355
Preheating 40 °C + OH to 80 °C	40 °C–80 °C OH	402.5
Preheating 60 °C + OH to 80 °C	60 °C–80 °C OH	497.5
Preheating 80 °C	80 °C	765

The obtained juices were evaluated regarding different analytical parameters. Colour measurement was performed directly in the fresh juice. For the determination of antioxidant activity (DPPH and ABTS method) and enzymatic activity such as peroxidase (POD) for both juices and polyphenol oxidase (PPO) for apple, juice samples were preserved in the frozen storage (−30 °C) until their use for the analysis.

2.3. Colour Measurements

Juice colour was measured using a Digieye colour measurement system (Verivide, UK). For each juice sample, L*, a* and b* parameters from CIELAB scale were measured. Total colour difference ΔE between untreated and treated juice samples and browning index (BI, for apple juice only) were calculated by equations (1) and (2), respectively. It has to be stated that the untreated juice showed a high degree of colour change due to oxidation and enzymatic browning. Hence, larger ΔE values, i.e., larger deviations from the untreated juice represent the preferred colour for high-quality juices.

$$\Delta E = \sqrt{\Delta L^2 + \Delta a^2 + \Delta b^2}, \tag{1}$$

$$BI = \frac{[(100(x - 0.31)]}{0.172} \tag{2}$$

where:

$$x = (a + 1.75L)/(5.645L + a - 3.012b) \tag{3}$$

The colour analyses were carried out in fifteen repetitions from each carrot and apple juice sample.

2.4. Antioxidant Activity (DPPH and ABTS Method)

The carrot and apple juices were centrifuged for 15 min at 10000 × g in a centrifuge (Eppendorf, Germany). The supernatants were for the antioxidant activity analysis by two methods, DPPH and ABTS.

The DPPH scavenging activity was evaluated according to [22]. A spectrophotometer (Photometer model U-1100, Hitachi, Ltd. Tokyo, Japan) set at wavelength of 517 nm was used to measure the absorbance. Quantification of the antioxidant activity was made by plotting a Trolox calibration curve ($r^2 = 0.9880$), and the results were expressed as mmol Trolox/L of juice.

The ABTS+• scavenging activity was carried out as described by [23]. The absorbance was measured with a spectrophotometer (Photometer model U-1100, Hitachi, Ltd. Tokyo, Japan) at 734 nm for a total time of 6 min. The results were expressed as mmol Trolox/L of juice by the quantification with Trolox standard curve ($r^2 = 0.9946$).

The values obtained are the average of three replicates from each juice sample.

2.5. Enzyme Activity

2.5.1. POD Assay

Carrot and apple juices were centrifuged at $10,000 \times g$ and 4 °C for 15 min. The supernatant was collected and analysed for POD activity at 470 nm and 25 °C, as described by [24]. The enzymatic extract was obtained by mixing 6.25 mL of juice sample with 12.5 mL of cold potassium phosphate buffer 0.1 M (pH 6.5) for 2 min. The POD substrate solution was prepared by mixing 0.1 mL of 99.5% of guaiacol, 0.1 mL of 30% of hydrogen peroxide and 99.8 mL of 0.1 M of potassium phosphate buffer (pH 6.5). POD activity was assessed by adding 150 µL of enzymatic extract to 3.35 mL of substrate solution in 10 mm pathlength glass cuvettes. POD activity for carrot and apple juice was calculated based on the slope (ΔA/min) of the linear portion of the plot of absorbance compared with time. An enzyme unit is defined as the enzyme activity that catalyses the conversion of 1 µmol of substrate into product in one minute.

2.5.2. PPO Assay

4-Methylcatechol 80 mM prepared in Mcllvaine's buffer solution at pH 7.5 was used as substrate, and 12.5 mL of cold Mcllvaine's buffer solution at pH 7.5 was added to 6.25 mL of enzymatic extract. PPO activity for apple juice was determined by reading the absorbance at 420 nm and 25 °C and calculated on the basis of the slope of the linear portion of the curve (ΔA/min). An enzyme unit is defined as the enzyme activity that catalyses the conversion of 1 µmol of substrate into product in one minute.

2.6. Data Analyses

The obtained data were analysed using parametric analysis of variance (ANOVA) using Tukey's HSD post-hoc test, performed with confidence level ($p < 0.05$). Conversely, when the normality of the distribution and the homogeneity of the variances were not satisfied, nonparametric ANOVA (Kruskal–Wallis) along with Holm's post-hoc tests were carried out ($p < 0.05$). All the statistical analyses were performed by using R statistical software R x64 3.4.3 (R foundation for statistical computing, Vienna, Austria).

3. Results and Discussion

3.1. Colour

Colour changes represent an indicator for enzymatic browning, as well as for process-induced browning due to heat-induced formation of Maillard products. Total colour variation (ΔE) between untreated and treated carrot and apple juice samples was analysed and is shown in Figure 1.

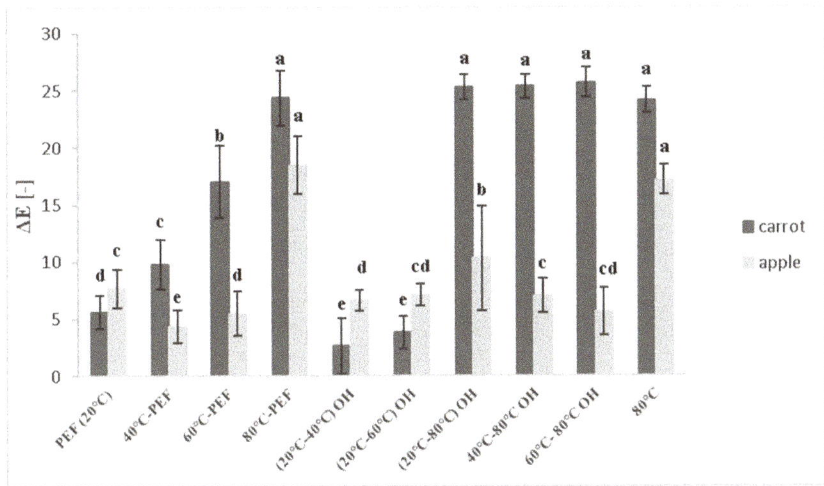

Figure 1. Total colour variation—ΔE between apple and carrot juices obtained from untreated and treated mash. Different letters indicate significant differences ($p < 0.05$) between samples.

Larger ΔE values represent a positive deviation from the untreated control sample that showed undesired browning due to enzyme activity and oxidation.

Juices from both raw materials pretreated at 80 °C with or without PEF showed higher ΔE values compared with those of control samples. According to the classification of [25], ΔE changes above 6 indicate great visible changes. The increase in ΔE reflects the increase in the lightness and decrease in the a* value of samples [26]. Since the untreated juice, which is considered as control sample in this case, showed unwanted browning and colour change due to enzyme activity and oxidation, higher ΔE values, i.e., higher deviation from the control juice colour indicated beneficial quality.

Lower ΔE values were observed for mash pretreated with PEF at room temperature and at 40 and 60 °C with OH.

The lowest ΔE values (4.36–5.49) were observed for preheated juice samples at 40 °C and 60 °C coupled with PEF, and the highest total colour differences were observed for samples preheated to 80 °C with or without additional PEF treatment (18.49 and 17.15, respectively).

The detected ΔE values between untreated and treated samples were even more pronounced for carrot compared with those for apple juice. In general, for both juices, higher L* values promoted also higher total colour differences compared with those of the control one. The browning index (BI) is a common parameter to describe colour and juice quality for apple. A decrease of the BI was found for apple samples preheated at 80 °C, coupled with or without PEF or OH treatment, in which was observed to reach BI from 115 to 119, compared with the untreated juice with much higher values of 142 (Figure 2).

Bhat et al. [18] reported similar results for bottle gourd treated with OH at 80 °C for 1 and 2 min (BI of 111 and 101, respectively). The progressive decrease in BI with increasing treatment temperature in apple mash indicates the relevance of enzymatic browning in untreated samples and the role of the temperature during mash treatment for the avoidance of unwanted reactions.

The main groups of pigments that are responsible for the characteristic colours in fruits and vegetables are carotenes and carotenoids, anthocyanins, chlorophylls and phenolic compounds. The main enzymes involved in biochemical degradations of plant compounds are peroxidase and polyphenoloxidase [6]. Moreover, another main cause of brown colour formation is nonenzymatic browning occurring in vegetable and fruit products. However, in the current study, the benefit from short-time thermal treatment of the juice of up to 80 °C for the inactivation of oxidative enzymes was more pronounced than the occurrence of detrimental colour changes due to nonenzymatic browning.

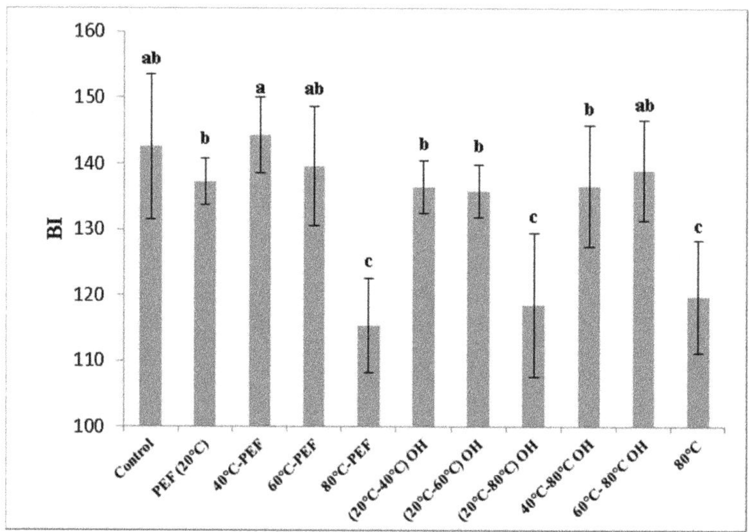

Figure 2. Browning index—BI in apple juice obtained from pretreated apple mash. Different letters indicate significant differences ($p < 0.05$) between samples. pulsed electric field (PEF); ohmic heating (OH).

3.2. Antioxidant Activity (DPPH and ABTS Method)

Figures 3 and 4 report the results of antioxidant activity, obtained with DPPH and ABTS antiradical activity methods, of differently obtained carrot and apple juices, respectively.

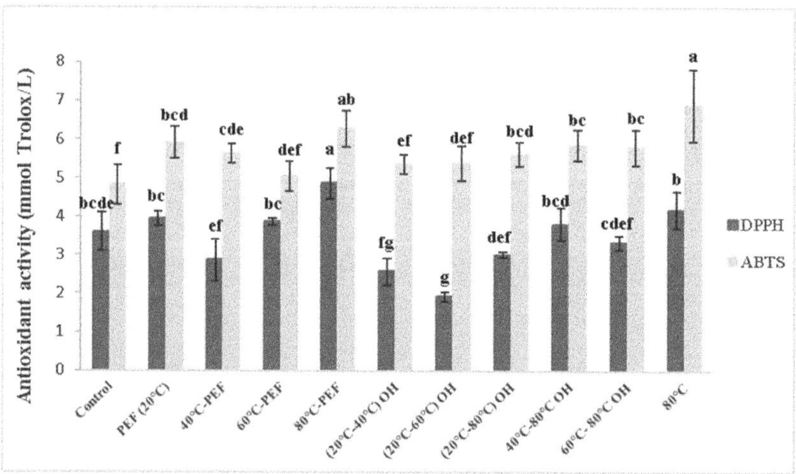

Figure 3. Antioxidant activity (DPPH and ABTS method) of carrot juices obtained from pretreated mash. Different letters indicate significant differences ($p < 0.05$) between samples.

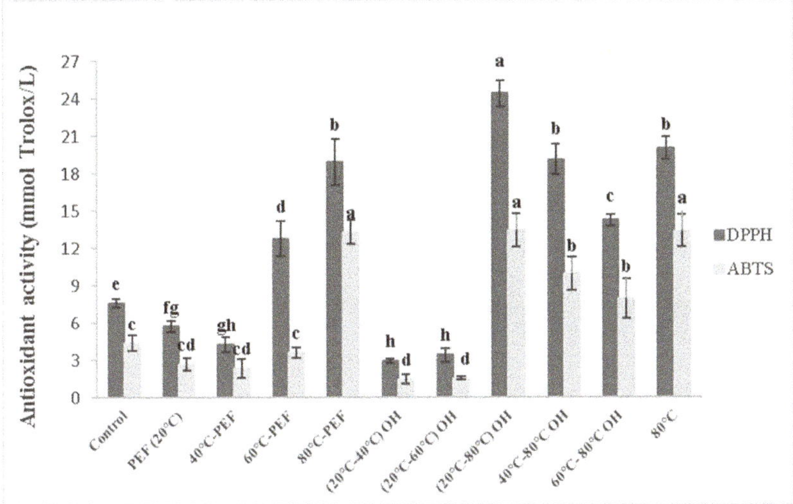

Figure 4. Antioxidant activity (DPPH and ABTS method) of apple juices obtained from pretreated mash. Different letters indicate significant differences ($p < 0.05$) between samples.

For carrot juice (Figure 3), a significantly higher antioxidant activity was obtained for carrot juices preheated to 80 °C with or without additional PEF or OH treatment with the ABTS method. Higher retention of bioactive compounds with DPPH method was observed for carrot mash pretreated at 80 °C coupled with PEF treatment. However, detected with DPPH method, the application of OH treatment reaching 40 and 60 °C reduced the antioxidant activity, in comparison with that of juice from the untreated control carrot mash. With ABTS method, no significant difference was found (Figure 3).

Significantly higher antioxidant activity, detected with both DPPH and ABTS methods, was obtained for apple juices preheated to 80 °C with or without additional PEF or OH treatment (Figure 4). Instead, OH reaching 40 °C and 60 °C reduced the antioxidant activity, with both used method, compared with that of the apple control juices, which might be due to the activation of degradative enzymes, such as peroxidase and polyphenoloxidase that lead to bioactive compounds' oxidative degradation.

Fruits and vegetables are good sources of natural antioxidants, containing carotenoids, vitamins, phenolic compounds, flavonoids, dietary glutathione and endogenous metabolites. However, the majority of the antioxidant activity of fruits and vegetables is derived from phenolic compounds (hydroxycinnamic acids, flavan-3-ols, anthocyanidins, flavonols and dihydrochalcones) rather than vitamin C and E, or β-carotene, due to their stronger activity against peroxil [27]. The peroxidase and polyphenoloxidase enzymes lead to the degradation of phenolic compounds and a subsequent loss of nutritional and sensorial values such as browning and off-flavour [28]. Moreover, heat treatment may influence the binding properties of bioactive compounds causing their higher release, but at the same time increase enzymatic or nonenzymatic degradation processes that can cause subsequent negative effects on quality of processed products [15].

In fact, higher temperature leads to the inactivation of the oxidative enzymes, thus reducing degradation effects and resulting in higher antioxidant activity in the juice. Moreover, additional effects other than those from thermal treatment need to be taken into account, since the electropermeabilization, induced by PEF and OH treatment, may contribute to the increased release of antioxidant compounds.

The detected difference between the two different methods used could be due to the fact that DPPH method is more sensitive to detect flavanones, while ABTS for the radical scavengers such as vitamin C [29].

For carrot and apple juice, PEF treatment without preheating did also not affect the extractability of bioactive compounds, which is in accordance with Shilling et al. [15], who reported no significant differences on total antioxidant activity between control and PEF-treated apple mash for different electric fields (1, 3 and 5 kV/cm).

3.3. Enzyme Activity

3.3.1. POD Activity

Process pretreatment for the juice production is an important operation in order to improve the quality of the vegetable and fruit raw materials as well as to avoid the activation of degradative enzymes such as POD and PPO that consequently provoke pigments and nutrients loss [30].

Peroxidase (POD) activity for differently obtained carrot juice is shown in Figure 5.

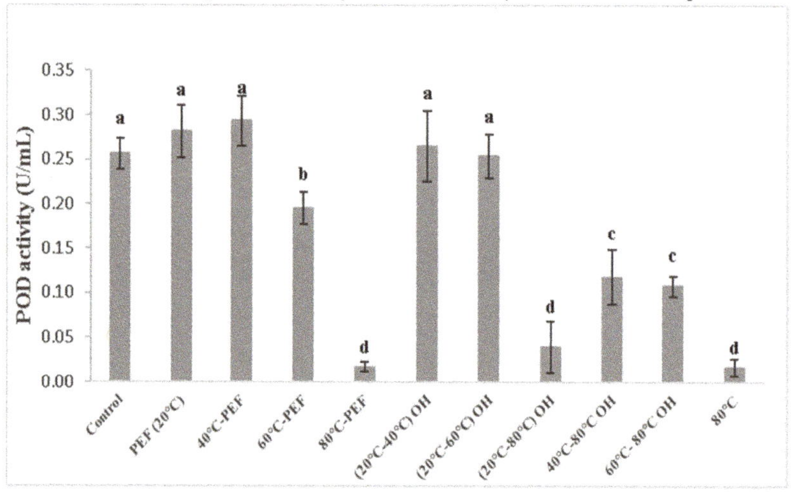

Figure 5. Peroxidase (POD) activity in carrot juice obtained from pretreated mash. Different letters indicate significant differences ($p < 0.05$) between samples.

This study revealed that the effect of only PEF treatment at 20 °C and 40 °C preheating could not reduce the activity of POD in both carrot (Figure 5) and apple juice [26], while for the samples preheated at 60°C, the reduction of POD activity was observed.

The highest POD inactivation could be reached by preheating to 80 °C with and without additional PEF or OH treatment. All pretreatment conditions with temperature at or up to 60 °C and 40–80 °C by OH treatment led to a decrease in POD activity in the carrot juice, while for apple juices, a greater reduction of the activity (from 50% to 90%) was achieved by 60 and 80 °C preheating temperatures with and without additional PEF or OH application. Enzyme inactivation in the juice, after higher PEF treatment intensities, for microbial inactivation and preservation purposes, is mainly related to secondary effects such as local temperature distributions, electrochemical reactions or formation of free radicals, instead of primary effects of electric field. For the treatment of mash, the PEF treatment intensity can be considered 10-fold lower and having no direct effect on fruit and vegetable mash ingredients.

POD activity decreased with increasing temperature, and almost no POD activity was detected in juice extracted when the treatment temperature reached 80 °C, especially in apple juice samples. High temperature leads to an increase in the internal energy of the enzymes, thus consequently causes the break of bonds that determine the three-dimensional structure of enzymes [18].

Moreover, with the increasing of temperature, the enzyme activity decreased and required a particular temperature–time combination for complete inactivation. Inadequate temperature led to a decrease in the enzyme activity time rather than complete inactivation, which may cause browning effect. In fact, Bhat et al. [18] reported similar results for OH-treated bottle gourd juice, where the temperature of 60 and 70 °C seemed to be not enough for complete enzyme inactivation, which instead was observed at 80 °C for 4 min.

Icier et al. [31] showed that OH treatment could be used for POD inactivation on pea puree at the range of 30–50 V/cm combined with the water blanching. Elez-Martínez et al. [32] reported a completely POD deactivation in orange juice after the application of PEF treatment at 35 kV/cm for 1500 μs.

Moreover, the variations of colour for carrot and apple juice pretreated with 80 °C with both PEF and OH applications could be explained by the decrease of enzyme activity, in fact, a correlation between colour and POD activity was found to be 0.8902 and 0.5166, respectively, for carrot and apple juice.

3.3.2. PPO Activity

Polyphenoloxidase (PPO) activity of differently obtained apple juice samples is shown in Figure 6.

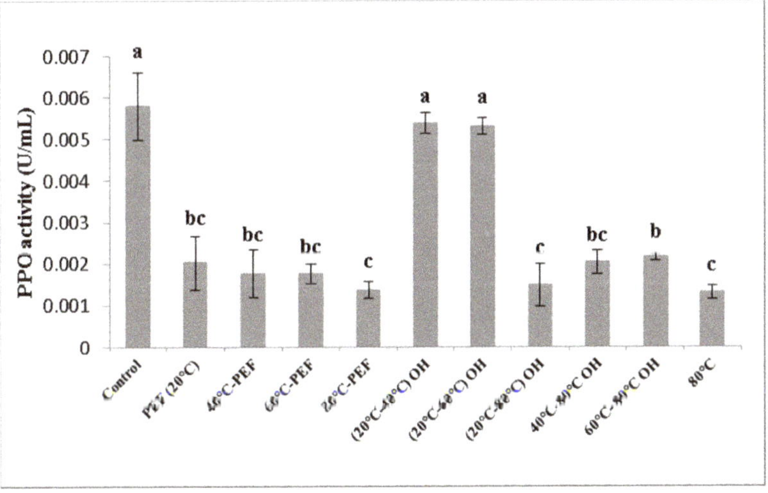

Figure 6. Polyphenoloxidase (PPO) activity of apple juice obtained from pretreated mash. Different letters indicate significant differences ($p < 0.05$) between samples.

PPO is an oxidoreductase enzyme which catalyses the oxidation of phenolic compounds in o-quinones, which are subsequently polymerized into brown pigments [33]. Heating treatment seems to be the most effective applied treatment for the stabilization of food products against microbial and enzyme activity. Nevertheless, thermal treatment has been shown to cause negative effects on quality and related nutritional compounds [9]. The mechanism of enzyme inactivation is not completely clear. Current results show empirical proof of protein modification by electrical fields [34] that may provoke a deformation or structural change of a protein, due to the interaction between the external electric field and the functional groups of the protein that allow its unfolding [32].

PPO activity was significantly decreased by PEF application at room temperature compared with that of the untreated control sample. In addition, a greater inactivation was achieved when the treatment temperature reached 80 °C, as well as with OH treatment and just preheating. Moreover, PPO inactivation was even more effective when a combination of temperatures and PEF or OH applications were used.

Similar results were reported by Turk et al. [35]. PPO activity was reduced in apple cider mash pretreated with PEF at 1 kV/cm for 100 µs. The loss of PPO activity could be explained also by the inhibition of the enzyme with oxidised phenolic compounds, in particular procyanidins [36].

Previous work reported similar results for PPO deactivation; the residual PPO activity was 35% after 14 min with OH treatment at 70 °C, by applying 35 V/cm in grape juice [30].

Liang et al. [37] found a significant decrease (33%) in PPO activity in freshly squeezed apple juice when preheated at 50 °C and treated with PEF at 27 kV/cm for 58.7 µs.

Saxena et al. [20] found a reduction of PPO activity up to 97.8 % by applying 32 V/cm with OH treatment at 90 °C for 5 min in sugarcane juice. Moreover, a greater increase in residual PPO activity was visible at 90 °C by increasing the holding times of OH treatment (5, 10, 15 and 20 min). The increase of the enzyme activity with the holding time at constant temperature was attributed to the pulsating OH treatment that may cause biochemical reactions by changing the molecular spacing and may result in a better interaction between substrate and enzyme [38]. A recent review on the impact of electric fields on enzymes is provided by [39].

4. Conclusions

Obtained results emphasize the role of thermal treatment for the inactivation of enzymes, reflected by improved colour values for juices exposed to 80 °C, independent of the PEF or OH application. The inactivation of POD and PPO was more pronounced when a temperature of 80 °C was achieved for both carrot and apple mash (around 90%).

However, a better retention of plant secondary metabolites from carrot and apple mashes could be achieved by additional PEF or OH application. PEF treatment was found to improve the release of such compounds, whereas OH contributed to a very fast volumetric heating that reduced the overall thermal load that the sample was exposed to. Based on the results, a combination of thermal and electric field pretreatments is required for the controlled release, inactivation and retention of ingredients. Thermal effects contributing to the colour, bioactive compounds retention and enzyme inactivation were found to be still important when applying nonthermal cell disintegration techniques such as PEF. However, both electrotechnologies, PEF and OH were found to positively contribute to improved juice quality by enhanced ingredient release and retention.

Author Contributions: C.M., T.F. and H.J. conceived and designed the experiments; C.M. and K.R. performed the experiments; C.M., K.R. and U.T. analysed the data; H.J. contributed reagents/ materials/ analysis tools; S.R. as Supervisor; C.M. contributed Writing-Original Draft Preparation; C.M., T.F., U.T., S.R., M.D.R. and H.J. contributed Writing-Review and Editing.

Funding: This work was partly supported by the Austrian Research Promotion Agency (FFG Project No. 859077).

Acknowledgments: Part of the equipment used in this study was financed by EQ-BOKU VIBT GmbH and belongs to the Center for Preservation and Aseptic Processing. Cinzia Mannozzi acknowledges the Marco Polo Programme (University of Bologna) for the travel grant.

Conflicts of Interest: The authors declare no conflict of interest.

References

1. Fincan, M.; Dejmek, P. In situ visualization of the effect of a pulsed electric field on plant tissue. *J. Food Eng.* **2002**, *55*, 223–230. [CrossRef]
2. Toepfl, S.; Mathys, A.; Heinz, V.; Knorr, D. Review: Potential of emerging technologies for energy efficient and environmentally friendly food processing. *Food Rev. Int.* **2006**, *22*, 405–423. [CrossRef]
3. Jaeger, H.; Schulz, M.; Lu, P.; Knorr, D. Adjustment of milling, mash electroporation and pressing for the development of a PEF assisted juice production in industrial scale. *Innov. Food Sci. Emerg. Technol.* **2012**, *14*, 46–60. [CrossRef]

4. Moussa-Ayoub, T.E.; Jaeger, H.; Youssef, K.; Knorr, D.; El-Samahy, S.; Kroh, L.W.; Rohn, S. Technological characteristics and selected bioactive compounds of *Opuntia dillenii cactus* fruit juice following the impact of pulsed electric field pre-treatment. *Food Chem.* **2016**, *210*, 249–261. [CrossRef] [PubMed]

5. Jaeger, H.; Roth, A.; Toepfl, S.; Holzhauser, T.; Engel, K.H.; Knorr, D.; Vogel, R.F.; Bandick, N.; Kulling, S.; Volker, H. Opinion on the use of ohmic heating for the treatment of foods. *Trends Food Sci. Technol.* **2016**, *55*, 84–97. [CrossRef]

6. Terefe, N.S.; Buckow, R.; Versteeg, C. Quality-related enzymes in fruit and vegetable products: Effects of novel food processing technologies, part 1: High-pressure processing. *Crit. Rev. Food Sci. Nutr.* **2014**, *54*, 24–63. [CrossRef] [PubMed]

7. Hendrickx, M.; Ludikhuyze, L.; Van den Broeck, I.; Weemaes, C. Effects of high pressure on enzymes related to food quality. *Trends Food Sci. Technol.* **1998**, *9*, 197–203. [CrossRef]

8. Jayaraman, K.S.; Ramanuja, M.N.; Dhakne, Y.S.; Vijayaraghavan, P.K. Enzymatic browning in some banana varieties as related to polyphenoloxidase activity and other endogenous factors. *J. Food Sci. Technol.* **1982**, *19*, 181–186.

9. Barsotti, L.; Dumay, E.; Mu, T.H.; Diaz, M.D.F.; Cheftel, J.C. Effect of high voltage electric pulses on protein-based food costituents and structutres. *Trends Food Sci. Technol.* **2001**, *12*, 136–144. [CrossRef]

10. Lasekan, O.; Ng, S.; Azeez, S.; Shittu, R.; Teoh, L.; Gholivand, S. Effect of Pulsed Electric Field Processing on Flavor and Color of Liquid Foods. *J. Food Proc. Preserv.* **2017**, *41*. [CrossRef]

11. Surowsky, B.; Fischer, A.; Schlueter, O.; Knorr, D. Cold plasma effects on enzyme activity in a model food system. *Innov. Food Sci. Emerg. Technol.* **2013**, *19*, 146–152. [CrossRef]

12. Sweiggert, R.M.; Carle, R. Carotenoid deposition in plant and animal foods and its impact on bioavailability. *Crit. Rev. Food Sci. Nutr.* **2017**, *57*, 1807–1830. [CrossRef] [PubMed]

13. Sanoner, P.; Guyot, S.; Marnet, N.; Molle, D.; Drilleau, J.F. Polyphenol profiles of French cider apple varieties (*Malus domestica* sp.). *J. Agric. Food Chem.* **1999**, *47*, 4847–4853. [CrossRef] [PubMed]

14. Khanizadeh, S.; Tsao, R.; Rekika, D.; Yang, R.; Charles, M.T.; Ruspasinghe, H.P.V. Polyphenol composition and total antioxidant capacity of selected apple genotypes for processing. *J. Food Comp. Anal.* **2008**, *21*, 396–401. [CrossRef]

15. Shilling, S.; Alber, T.; Toepfl, S.; Neidhart, S.; Knorr, D.; Schieber, A.; Carle, R. Effects of pulsed electric field treatment of apple mash on juice yield and quality attributes of apple juices. *Innov. Food Sci. Emerg. Technol.* **2007**, *8*, 127–134. [CrossRef]

16. Caminiti, I.M.; Noci, F.; Muñoz, A.; Whyte, P.; Morgan, D.J.; Cronin, D.A.; Lyng, J.G. Impact of selected combinations of non-thermal processing technologies on the quality of an apple and cranberry juice blend. *Food Chem.* **2011**, *124*, 1387–1392. [CrossRef]

17. Yan, L.G.; He, L.; Xi, J. High intensity pulsed electric field as an innovative technique for extraction of bioactive compound. A review. *Crit. Rev. Food Sci. Nutr.* **2017**, *57*, 2877–2888. [CrossRef]

18. Bhat, S.; Saini, C.S.; Sharma, H.K. Changes in total phenolic content and color of bottle gourd (*Lagenaria siceraria*) juice upon conventional and ohmic blanching. *Food Sci. Biotech.* **2017**, *26*, 29–36. [CrossRef]

19. Saberian, H.; Hamidi-Esfahani, Z.; Ahmadi Gavlighi, H.; Banakar, A.; Barzegar, M. The potential of ohmic heating for pectin extraction from orange waste. *J. Food Proc. Preserv.* **2018**, *42*, e13458. [CrossRef]

20. Saxena, J.; Ahmad Makroo, H.; Srivastava, B. Effect of ohmic heating on Polyphenol Oxidase (PPO) inactivation and color change in sugarcane juice. *J. Food Proc. Eng.* **2017**, *40*, 1–11. [CrossRef]

21. Mannozzi, C.; Fauster, T.; Haas, K.; Tylewicz, U.; Romani, S.; Dalla Rosa, M.; Jaeger, H. Role of thermal and electric field effects during the pre-treatment of fruit and vegetable mash by pulsed electric fields (PEF) and ohmic heating (OH). *Innov. Food Sci. Emerg. Technol.* **2018**, *48*, 131–137. [CrossRef]

22. Amarowicz, R.; Naczk, M.; Shahidi, F. Antioxidant activity of various fractions of non-tannin phenolics of canola hulls. *J. Agric. Food Chem.* **2000**, *48*, 2755–2759. [CrossRef] [PubMed]

23. Re, R.; Pellegrini, N.; Proteggente, A.; Pannala, A.; Yang, M.; Rice-Evans, C. Antioxidant activity applying an improved ABTS radical cation decolorization assay. *Free Radic. Biol. Med.* **1999**, *26*, 1231–1237. [CrossRef]

24. Morales-Blancas, E.F.; Chandia, V.E.; Cisneros-Zevallos, L. Thermal inactivation kinetics of peroxidase and lipoxygenase from broccoli, green asparagus and carrots. *J. Food Sci.* **2002**, *67*, 146–154. [CrossRef]

25. Cserhalmi, Z.; Sass-Kiss, A.; Tóth-Markus, M.; Lechner, N. Study of pulsed electric field treated citrus juices. *Innov. Food Sci. Emerg. Technol.* **2006**, *7*, 49–54. [CrossRef]

26. Mannozzi, C.; University of Bologna, Cesena, Italy. Unpublished work. 2017.

27. Zhang, D.; Hamauzu, Y. Phenolic compounds and their antioxidant properties in different tissues of carrots (*Daucus carota L.*). *J. Food Agric. Environ.* **2004**, *2*, 95–100.

28. Vámos-Vigyázó, L. Prevention of enzymatic browning in fruits and vegetables: A review of principles and practice. *ACS Symp. Ser.* **1995**, *600*. [CrossRef]

29. Del Caro, A.; Piga, A.; Vacca, V.; Agabbio, M. Changes of flavonoids, vitamin C and antioxidant capacity in minimally processed citrus segments and juices during storage. *Food Chem.* **2004**, *84*, 99–105. [CrossRef]

30. Gonçalves, E.M.; Pinheiro, J.; Abreu, M.; Brandão, T.R.S.; Silva, C.L. Carrot (*Daucus carota L.*) peroxidase inactivation, phenolic content and physical changes kinetics due to blanching. *J. Food Eng.* **2010**, *97*, 574–581. [CrossRef]

31. Icier, F.; Ilicali, C. The effects of concentration on electrical conductivity of orange juice concentrates during ohmic heating. *Eur. Food Res. Technol.* **2005**, *220*, 406–414. [CrossRef]

32. Elez-Martínez, P.; Soliva-Fortuny, R.C.; Martín-Belloso, O. Comparative study on shelf life of orange juice processed by high intensity pulsed electric fields or heat treatment. *Euro. Food Res. Technol.* **2006**, *222*, 321. [CrossRef]

33. Murata, M.; Tsurutani, M.; Tomita, M.; Homma, S.; Kaneko, K. Relationship between apple ripening and browning: Changes in polyphenol content and polyphenol oxidase. *J. Agric. Food Chem.* **1995**, *43*, 1115–1121. [CrossRef]

34. Freedman, K.J.; Haq, S.R.; Edel, J.B.; Jemth, P.; Kim, M.J. Single molecule unfolding and stretching of protein domains inside a solid-state nanopore by electric field. *Sci. Rep.* **2013**, *3*, 1638. [CrossRef] [PubMed]

35. Turk, M.F.; Billaud, C.; Vorobiev, E.; Baron, A. Continuous pulsed electric field treatment of French cider apple and juice expression on the pilot scale belt press. *Innov. Food Sci. Emerg. Technol.* **2012**, *14*, 61–69. [CrossRef]

36. Le Bourvellec, C.; Le Quéré, J.M.; Sanoner, P.; Drilleau, J.F.; Guyot, S. Inhibition of apple polyphenol oxidase activity by procyanidins and polyphenol oxidation products. *J. Agric. Food Chem.* **2004**, *52*, 122–130. [CrossRef] [PubMed]

37. Liang, Z.; Cheng, Z.; Mittal, G.S. Inactivation of spoilage microorganisms in apple cider using a comntinuous flow pulsed electric field system. *LWT Food Sci. Technol.* **2006**, *39*, 351–357. [CrossRef]

38. Castro, I.; Macedo, B.; Teixeira, J.A.; Vicente, A.A. The effect of electric field on important food-processing enzymes: Comparison of inactivation kinetics under conventional and ohmic heating. *J. Food Sci.* **2004**, *69*, 696–701. [CrossRef]

39. Poojary, M.M.; Roohinejad, S.; Koubaa, M.; Barba, F.J.; Passamonti, P.; Jambrak, A.R.; Oey, I.; Greiner, R. Impact of Pulsed Electric Fields on Enzymes. In *Handbook of Electroporation*; Miklavčič, D., Ed.; Springer: Cham, Switzerland, 2016. [CrossRef]

Article

The Influence of Chemical Structure and the Presence of Ascorbic Acid on Anthocyanins Stability and Spectral Properties in Purified Model Systems

Rachel Levy, Zoya Okun and Avi Shpigelman *

Faculty of Biotechnology & Food Engineering, Russell Berrie Nanotechnology Institute, Technion, Israel Institute of Technology, Haifa 3200003, Israel; rachel447@campus.technion.ac.il (R.L.); zoya@bfe.technion.ac.il (Z.O.)
* Correspondence: avis@bfe.technion.ac.il; Tel.: +97-27-7887-1867

Received: 24 May 2019; Accepted: 10 June 2019; Published: 12 June 2019

Abstract: The loss of color pigment is an important quality factor of food products. This work aimed to systematically study, in purified model systems, the influence of anthocyanins' structure (by increasing the size of the conjugated sugar) and the presence of ascorbic acid on their stability and spectral properties during storage at two pH levels relevant to medium and high acid foods (6.5 and 4.5, respectively). Anthocyanins (cyanidin (Cy), cyanidin 3-O-β-glucoside (Cy3G) and cyanidin 3-O-β-rutinoside (Cy3R)) displayed first-order degradation rates, presenting higher stability in acidic medium and enhanced stability with increasing size of conjugated sugar. The addition of ascorbic acid resulted in significantly enhanced degradation. Changes in ultra violet visible (UV-VIS) spectral properties presented a decrease in typical color intensity and pointed towards formation of degradation products. Identification and kinetics of formation for cyanidin degradation products were obtained by high performance liquid chromatography system-mass spectrometry (HPLC-MS).

Keywords: anthocyanins; ascorbic acid; UV-Vis; HPLC-MS; kinetics; shelf life

---

1. Introduction

Color is an important sensory characteristic that consumers rely on when making food choices, and even the slightest changes due to processing or during shelf life can deter a potential buyer and alter the consumer's perception of product quality. The loss of color pigment is an important shelf life factor for many food products, especially for those undergoing thermal treatment [1]. Anthocyanins are a large group of natural water-soluble pigments that are responsible for red, blue and purple colors in many fruits, vegetables, flowers, leaves, stems and roots [2,3], and their deterioration is an important factor in loss of color in many food products [1]. The bright colors of the pigments are due to the conjugated double bonds that are responsible for the absorption of visible light [4], and the pH dependent equilibrium allows their utilization as natural food colorants [3,5]. These pigments usually occur in the glycosylated polyhydroxy and polymethoxy form of 2-phenylbenzopyrylium salts (flavylium ion skeleton). Beyond their effect on color, the increasing interest in the research of these molecules stems from their reported various health promoting properties such as anti-inflammatory [6], anti-cancer [7] anti-cardiovascular [8] and other bioactivities [8–11]. As members of the polyphenol group, anthocyanins and anthocyanidins (the aglycone form of anthocyanins) possess antioxidant properties [8,9], the conjugated sugar moiety of the anthocyanins is known to reduce the radical scavenging activity as compared to aglycone [8,12]. It was suggested that attached sugar unit reduces the ability of the anthocyanin radical to delocalize electrons [8].

While anthocyanins are widely occurring in food products, they tend to be unstable during processing and storage and to be degraded and/or decolorized. Some studies explore the chemistry and stability of anthocyanins, mostly in fruit and vegetables and their processed products, but also in some model systems [9,12–16]. The stability is reported to be affected by pH, temperature, chemical structure of the pigment, concentration, solvents, oxygen, light, enzymes, etc. [2,3,9,17,18]. During processing and storage, anthocyanins degradation increases with the rise in temperature [13,19]. At room temperature, color is reported to be stable only in acidic media. In alkaline media, cleavage of the pyrylium ring takes place and decrease in color intensity is noticed [20]. It might be assumed that thermal degradation of anthocyanins begins in the opening of the heterocycle and the formation of the chalcone form [21,22]. Heating shifts the equilibrium towards the chalcone and the reversion of chalcone to flavylium is slow to impossible [22]. Thermal degradation of anthocyanins follows first-order reaction kinetics [23–25] and can be reduced by decreasing the pH [1,25]. Another suggested mechanism excludes the formation of chalcone glycoside form, and by combination of heat and pH levels (2–4) hydrolysis of the glycosidic bond occurs, followed by conversion of the aglycone to chalcone and then to degradation products [22]. At aqueous solution at pH 2–4, temperature elevation leads to the hydrolysis of glycosidic bond, resulting in the loss of sugar moieties of the anthocyanins, leading to further loss of color as the anthocyanidins are much less stable than anthocyanins [20]. The presence of the sugar group is responsible for increased water solubility and stability (compared to aglycone) [4,26], with the number of sugar rings also suggested to influence the stability [22,27,28].

Many food products, especially juices, are fortified with ascorbic acid (AA), a natural antioxidant, to protect against oxidation and to increase the nutritional value [29,30]. AA might have several negative influences on anthocyanins' stability. In the presence of oxygen, AA can accelerate the degradation of anthocyanins and enhance the formation of polymer pigment, which results in anthocyanins pigment bleaching [29–31]. The exact mechanism is still controversial and addition of AA to anthocyanins results in increase of the degradation rate of both molecules. The postulated mechanisms are either direct condensation of AA with anthocyanins or formation of hydrogen peroxide and oxidative cleavage of the pyrylium ring by peroxide [26,29,31]. In previous works, the focus of anthocyanins stability in the presence of AA is explored in food matrixes [32,33] or with a focus on specific molecules [29,31].

The aim of this study was to explore, systematically, in purified model systems, the influence of anthocyanins' structure (by increasing the size of the conjugated sugar), pH (6.5 and 4.5) and the presence of ascorbic acid on their stability and spectral properties during simulated shelf life. The degradation was tested in purified model systems to avoid possible interferences from food matrix. In addition, we aimed to better understand the kinetics of the formation of degradation products for the most sensitive of the tested anthocyanins—cyanidin. To our knowledge, no other work examines the stability of series of three molecules that differ only by the presence and type of sugar moiety, by high performance liquid chromatography system-mass spectrometry (HPLC-MS) analysis and quantification, monitoring of the spectral properties and identification of the degradation products in purified system.

2. Materials and Methods

2.1. Materials

Cyanidin chloride (Cy) was purchased from Tokiwa phytochemical Co., LTD (Tokyo, Japan) (CAS number: 528-58-5). Cyanidin 3-O-β-glucopyranoside (CAS number: 7084-24-4) and Cyanidin 3-O-(6''-O-α-rhamnopyranosyl-β-glucopyranoside) (CAS number: 18719-76-1) were purchased from Polyphenols AS (Sandnes, Norway). L-(+)-Ascorbic acid 99+%, was purchased from Alfa Aesar (Heysham, England) (CAS number: 50-81-7). 2,4,6-trihydroxybenzaldehyde (CAS number: 487-70-7) and 2,4,6-trihydroxybenzoic acid (CAS number: 71989-93-0) were purchased from Sigma-Aldrich (Saint Louis, MO, USA). 3,4-dihydroxybenzoic acid was purchased from Fluka Chemika (Buchs, Switzerland) (CAS number: 99-50-3).

2.2. Methods

2.2.1. Buffered and Stock Solutions

Stock solutions of Cy3G, Cy3R, 2,4,6-trihydroxybenzaldehyde, 2,4,6-trihydroxybenzoic acid and 3,4-dihydroxybenzoic acid (2.5 mM) were prepared in pure methanol. Cy stock solution (2.5 mM) was prepared in methanol and 1% (*v/v*) formic acid. All stock solutions were covered with aluminum foil and stored at −40 °C. Acetate buffered solution (20 mM, pH = 4.5) was prepared using acetic acid and sodium acetate. Phosphate buffered solutions (20 mM, pH = 1.5 and pH = 2.5) were prepared using phosphoric acid. Phosphate buffered solution (20 mM, pH = 6.5) was prepared using sodium phosphate, monobasic and dibasic.

2.2.2. Preparation of Anthocyanins Model Solutions

The stability of Cy, Cy3G and Cy3R was conducted by diluting the Cy, Cy3G, and Cy3R stock solutions in phosphate buffered solution pH = 6.5 and acetate buffered solution pH = 4.5 for final concentration of 0.1 mM (4% methanol). For exploring the influence of AA, similar solutions were prepared with an addition of AA (200 mg/L) the pH of the samples was adjusted correspondingly. Appropriate controls of AA without anthocyanins and pure buffered solutions were prepared. The samples were vortexed and stored at least in duplicate in the dark at 15, 23 and 37 °C. Immediately after preparation ($t = 0$) and after 6 h, 27 h, 32 h, 72 h and 146 h, the samples were diluted with acidic buffered solutions (pH = 1.5 for Cy and pH = 2 for Cy3G and Cy3R) for final concentration of 0.05 mM (2% methanol) to get the maximum stability of the anthocyanins for HPLC-MS analysis. The samples when then frozen and stored at −40 °C until analysis.

2.2.3. Quantification of the Model Solutions Degradation and Degradation Products

The acidified samples were filtered by a 0.45 µm PVDF syringe filters (Merck Millipore LTD., Carrigtwohill, Ireland) and injected to an Agilent 1260 high performance liquid chromatography system with mass spectrometry (HPLC-MS) 6120 Quadrupole (Santa Clara, CA, USA) equipped with InfinityLab Poroshell 120 EC-C18 column (2.1 × 150 mm, 1.9 micron) protected by guard column (Agilent, Santa Clara, CA, USA). The injection volume was 10 µL. The flow rate was 0.19 mL/min, column temperature was set to 30 °C and the sampler temperature was set to 4 °C. The mobile phase included: an aqueous solution of 1% formic acid (A) and an aqueous solution of 50% 1% formic acid, 25% methanol and 25% acetonitrile (B). A gradient of 25% B for 0–15 min, 100% B for 15–17 min, 25% B for 17–22 min and 25% B for 22–23 min was used. The chromatograms were acquired at 270 nm, to detect not only the anthocyanins and at 516 nm for anthocyanins quantification. All analysis in the HPLC-MS were carried out in full scan mode from 150 to 870 *m/z* API-ES source in negative and positive mode. The capillary: negative 3500 V, positive 4000 V, nebulizer gas (N2) 60 psi, dry gas (N2) 13 (1/min), dry temperature 350 °C. Validation of the Cy, Cy3G, and Cy3R peaks was performed by negative (286, 448, and 594) or positive (288, 450, and 596) selected ion measurement (SIM), respectively. Relative concentration of the remaining anthocyanins for each sample were calculated as the peak area of experimental samples divided by the average area of the same sample at $t = 0$ min at 516 nm. For the quantification of Cy degradation products, relative concentration of degradation products was calculated at 270 nm as the average peak area of experimental samples divided by the maximum average area of the samples over the experimental time.

2.2.4. Qualitative Spectroscopic Analysis of Model Solutions

For qualitative study of the color stability, all samples of model solution were loaded on a 96-well quartz microplate. The appropriate controls containing AA in buffered solutions without anthocyanins and pure buffered solutions were prepared and also loaded on the microplate. The samples' ultra violet visible (UV-VIS) absorbance was recorded as a function of time at the range of 250–600 nm using a Synergy H1 hybrid multi-mode reader (Biotek, Winooski, VT, USA) connected to a computer. The plate was stored in the dark at all the examined temperatures.

2.2.5. Calculation of Reaction Rate Constants

The first-order reaction rate constants (k) were calculated by nonlinear curve fitting using Origin 2018 (OriginLab, Northampton, MA, USA).

2.2.6. Statistical Analysis

The significance of the influence of chemical structure on stability was calculated using the nonlinear model to compare multiple databases by Origin 2018 (OriginLab), assessing if two datasets were significantly different from each other by an F-test ($p < 0.05$).

3. Results and Discussion

3.1. Degradation of Cyanidin

3.1.1. Stability of Stock Solution

One of our goals was to characterize the degradation of cyanidin by spectral changes and the formation of degradation products. Therefore, we first aimed to find the most stable conditions for preparation of Cy stock solution. From the tested solvents and storage conditions, only storage in methanol with 1% formic acid at −40 °C allowed conservation of more than 95% of Cy after 30 days. In all other tested stock solutions (pure methanol or methanol:water (20:80 and 50:50) with 1% formic acid, buffers at pH 2–7 stored at −20 °C and −40 °C), degradation of Cy was detected by HPLC-UV resulting in formation of new peaks.

3.1.2. Stability of Cyanidin and the Formation of Degradation Products

Anthocyanins are more stable in acidic solutions than in alkaline or neutral media [22]. In solution media, four main equilibria forms are known: flavylium cation, quinoidal base, carbinol pseudobase and chalcone (*cis* or *trans*). In acidic solutions up to pH 3, the main form is flavylium cation (red color). By raising the pH above 4, the color pigment and concentration of the flavylium cation decrease and the color can turn blue due to quinoidal base or even colorless or yellowish pigment due to chalcone and pseudobase. The process might be reversible until the point where the pH value is too high and unstable ionic chalcone is formed. At this stage, the regeneration of color cannot be achieved. The chalcone was suggested to further degrade to 2,4,6-trihydroxybenzaldehyde and phenolic compound or coumarin [3,34]. In many food products containing anthocyanin, the pH range is 2–4 resulting in the flavylium cation as the main specie [34,35].

The stability of Cy during 146 h at 37 °C and two pH values, 4.5 and 6.5, relevant to mild and low acid foods, was monitored by HPLC-UV-MS (Figure 1). The presented chromatograms in Figure 1A,C clearly shows the degradation of Cy (Rt = 15.9, peak number (4)) over time at both pH values. No significant difference between the degradation of Cy in pH = 4.5 compared to pH = 6.5 was noticed, possibly due to the quick degradation at both pH conditions. The chromatograms also clearly present, for the first time in non-highly acidic conditions, the kinetics of formation of the degradation products. Reported works identify possible degradation products of Cy as chalcone (cis or trans), 3,4-dihydroxybenzoic acid, 2,4,6-trihydroxybenzaldehyde and 2,4,6-trihydroxybenzoic acid (Figure 2) yet in pH values not relevant for food products [13,19]. In this study, three main degradation products were identified and monitored over time (Figure 1B,D): 3,4-dihydroxybenzoic acid (Rt = 3.8, peak number (1)), 2,4,6-trihydroxybenzaldehyde (Rt = 11.9, peak number (2)) and chalcone (cis or trans, Rt = 12.2, peak number (3)). The peaks of 3,4-dihydroxybenzoic acid and 2,4,6-trihydroxybenzaldehyde were compared to standards and identified by retention time (Rt), absorption spectra and mass spectral analysis. In both distinct pH levels, the same formed degradation products were observed and the general trend of their formation and further degradation was similar. At pH = 4.5 and pH = 6.5, the chalcones appeared up to 32 h and 6 h respectively, and then decreased with time. These results support previous suggested mechanism [13] showing that chalcones are being consumed in subsequent degradation process and are intermediate deterioration products. At the same time, the amount of the other two products continuously increased with time. Additional suggested secondary degradation product, 2,4,6-trihydroxybenzoic acid [13,19], was not identified in any of the systems during the examined time. After 146 h, a new peak (5) (Rt = 9.5) appeared at both pH values only at the chromatograms acquired at 270 nm with fragmentation of 289 (as the major mass in the mass spectrum), but it was not identified.

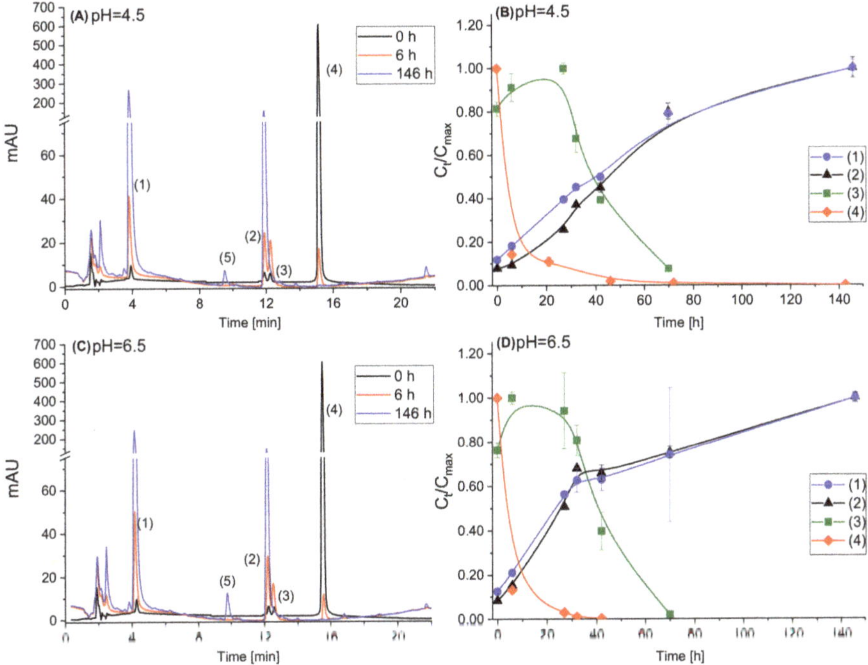

Figure 1. Stability of Cy and the formation of degradation products in buffered solutions stored at 37 °C: (**A**) a typical HPLC chromatogram at 270 nm of Cy solution (pH = 4.5) after 0, 6 and 146 h; (**B**) relative concentration (Ct/Cmax) of Cy and Cy degradation products (pH = 4.5); (**C**) a typical HPLC chromatogram at 270 nm of Cy solution (pH = 6.5) after 0, 6 and 146 h; and (**D**) relative concentration (Ct/Cmax) of Cy and Cy degradation products (pH = 6.5). Compounds identification: (1) 3,4-dihydroxybenzoic acid (Rt = 3.8); (2) 2,4,6-trihydroxybenzaldehyde (Rt = 11.9); (3) chalcone (Rt = 12.2); (4) Cy (Rt = 15.9); and (5) unidentified degradation product (Rt = 9.5). Quantification was made by HPLC-UV absorbance of the peak at 270 nm. Error bars represent standard error (*n* = 2). In some cases, they are smaller than the symbols. The lines in (**B,D**) are to guide the readers' eye.

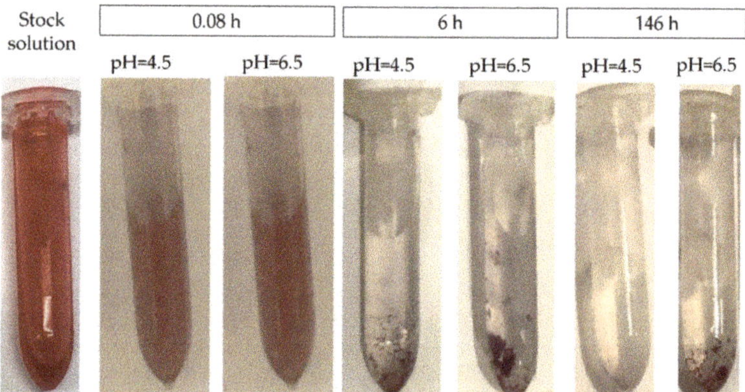

Figure 2. Chemical structure of anthocyanins and suggested degradation products: (**1**) 3,4-dihydroxy benzoic acid; (**2**) 2,4,6-trihydroxybenzaldehyde; (**3**) chalcone; (**4**) Cyanidin (Cy); (**6**) Cyanidin 3-O-β-glucoside (Cy3G); and (**7**) Cyanidin 3-O-β-rutinoside (Cy3R)). Structures were obtained using SciFinder® application.

The visual appearance of the Cy stock (2.5 mM) and Cy sample (0.1 mM) solutions over time are presented in Figure 3. Cy concentrations were below the reported maximal solubility of Cy (0.17 mM) which is the least soluble from the three studied molecules [36]. The color of Cy immediately changed from red into blue-purple at both pH values. After 6 h, the formation of insoluble purple sediment was observed at all studied temperatures (15 °C, 23 °C, and 37 °C). While chemical instability of anthocyanins is well documented, to the best of our knowledge, the formation of physical instability (sediment) during shelf-life/storage is not reported before. We suggest that the observed sediment is, at least partially, the outcome of chemical instability resulting in the formation of insoluble degradation products that would not be detected in the chromatograms in Figure 1A,C, although we cannot exclude involvement of Cy molecules themselves. The identification of the purple residues, the kinetics and factors affecting their formation should be further studied as the sediment may have a major impact on both sensorial and nutritional properties.

Figure 3. Visual color of Cy stock solution and Cy in buffered solution pH = 4.5 and pH = 6.5, stored at 37 °C after 5 min, 6 h, 42 h and 143 h.

3.2. The Influence of Chemical Structure of Anthocyanins on Stability and Spectral Properties

3.2.1. Stability of Stock Solutions and Buffered Solutions

Cy3G and Cy3R stock solutions were prepared in pure methanol and stored at −40 °C allowing the conservation of more than 98% of the original compounds after 30 days. In buffered solutions that were also tested for stock formation, the lowest degradation of these molecules was observed in phosphate buffered solution pH = 2.5.

3.2.2. The influence of the Sugar Moiety Size

To understand the influence of the size of the conjugated sugar moiety, the stability of Cy, Cy3G and Cy3R was monitored at different pH conditions (4.5 and 6.5) and temperatures (37 °C, 23 °C, and 15 °C). Figure 4 presents stability results as the relative concentration of the peak area divided by the initial peak area at 516 nm for the compounds stored at 37 °C. In addition, the changes in the spectral properties were monitored by collecting UV-VIS absorbance spectra over time.

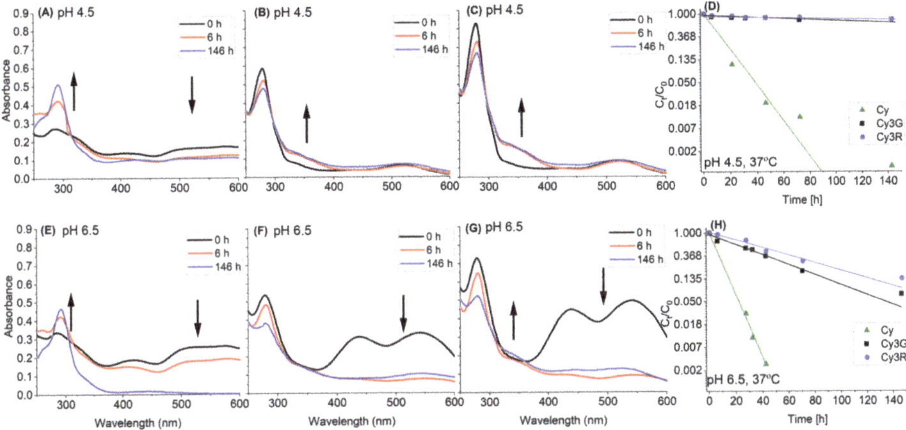

Figure 4. Stability (by HPLC) and changes in the absorbance spectrum of Cy, Cy3G and Cy3R stored at 37 °C. (**A**) average absorbance spectrum of Cy (pH = 4.5); (**B**) average absorbance spectrum of Cy3G (pH = 4.5); (**C**) average absorbance spectrum of Cy3R (pH = 4.5); (**D**) relative concentration (compared to $t = 0$) of Cy, Cy3G and Cy3R over time (pH = 4.5); (**E**) average absorbance spectrum of Cy (pH = 6.5); (**F**) average absorbance spectrum of Cy3G (pH = 6.5); (**G**) average absorbance spectrum of Cy3R (pH = 6.5); and (**H**) relative concentration (compared to $t = 0$) Cy, Cy3G and Cy3R over time (pH = 6.5). Quantification was made by HPLC-VIS absorbance of the peak at 516 nm and presented as percentage of the peak area divided by the initial peak area. (**D,H**) Error bars represent standard error ($n = 2$); in some cases, they are smaller than the symbols. The linear line represents fit to first-order degradation kinetics.

Cy degradation was immediate; Cy3G and Cy3R presented significantly higher stability than Cy at both pH values ($p < 0.05$) (Figure 4D,H). These results support the hypothesis that the sugar moiety stabilizes the molecules [20,22] and that the decrease in absorbance might be used as indicator for the formation of degradation products [13,26,31]. While the decrease in absorbance is often used as a simple method for quantification of anthocyanins degradation [14,28], at the higher pH, the correlation between concentrations quantified by HPLC and the decrease in the typical VIS absorbance above 500 nm (measured by spectrophotometer) seems less clear (Figure 4G,H), therefore should be verified and treated with caution. This could stem from the various intermediate reversible products that are likely to appear at higher pH levels. Beside the decrease in color intensity in the visible range, increase in the absorbance was observed in the UV range for all samples at pH 4.5 and in Cy solutions in pH 6.5

(Figure 4A–C,E). This is in good agreement with the degradation products detected in the HPLC-UV elution profile of all samples at 270 nm. When comparing the size of the sugar moiety, a significant difference (in the value of the slope, k) between the degradation of Cy3G and Cy3R in the two pH values ($p < 0.05$) was detected, although the effect was larger at pH 6.5 (Figure 4D,H). This result further presents that the disaccharide moiety stabilizes the anthocyanin more than the monosaccharide. In addition, at alkaline solutions, the decrease in stability of Cy3G and Cy3R, was faster ($p < 0.05$), as expected [22]. In contrast, there was no significant difference in the stability of Cy depending on the pH value, yet further work verifying if this is the outcome of extremely fast degradation at both pH levels or a mechanism that is not as pH depended (in these pH range) as in Cy3G and Cy3R is needed.

The complete anthocyanin thermal degradation mechanism is still unclear [21,27,37]. As mentioned, it was suggested before that thermal degradation of anthocyanins in aerobic conditions and buffer solution pH = 2–4 starts with the formation of chalcone glycoside followed by hydrolysis of the sugar moieties [21]. Another proposed mechanism suggested hydrolysis of the sugar moieties and then formation of chalcone [29]. However, after the formation of the chalcone, the degradation products of anthocyanins should eventually be the same as those of the aglycone [21]. In our HPLC-MS analysis of Cy3G and Cy3R solutions at both pH values (data not shown), a decrease in the peak areas of Cy3G and Cy3R was clearly observed, followed by the appearance of new peaks detectable at 270 nm. However, degradation products of Cy3G and Cy3R have not yet been fully characterized.

3.3. The Influence of the AA Addition on the Stability of Anthocyanins

To better understand the influence of AA addition to anthocyanins, the stability of Cy3G and Cy3R was measured during time at different pH conditions. The fact that this work focuses only on two isolated molecules without the presence of food matrix allows a better understanding of the effects of AA on anthocyanin stability. The concentration of AA used in the model system, 200 mg/L, was similar to that found in fruit juices [38]. Stability results are presented as relative (compared to $t = 0$) concentrations quantified as peak area at 516 nm at the two pH values (Figure 5).

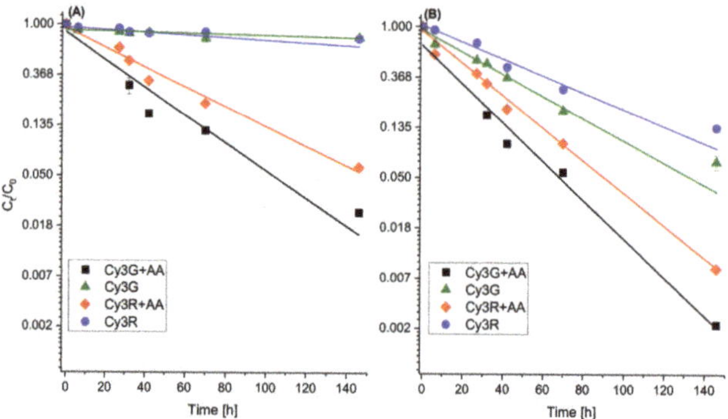

Figure 5. Stability of Cy3G and Cy3R in buffered solutions (pH 4.5 (**A**); and pH 6.5 (**B**)) over time with and without AA, stored at 37 °C. Quantification was made by HPLC absorbance of the peak at 516 nm and presented as relative concentration of the peak area divided by the initial peak area. Error bars represent standard error ($n = 2$). In most cases, they are smaller than the symbols.

It is known that AA in the presence of oxygen accelerates the decomposition of several anthocyanins and leads to bleaching of anthocyanin pigments [29–31,39]. Our results show that, after 6 h, there was a drastic decrease in the peak area of AA (by HPLC, areas are not shown) at 270 nm in the solution with anthocyanins compared to the control solution (only AA). The degradation mechanism of anthocyanins in the presence of AA is controversial. The postulated mechanisms are either direct condensation of AA with anthocyanins [40] or by free radical mechanism (formation of hydrogen peroxide and oxidative cleavage of the pyrylium ring by this peroxide) [26,31]. In all model systems, with and without the addition of AA, decrease in the peak areas of Cy3G and Cy3R were followed by appearance of new peaks in the HPLC chromatogram. Difference in the HPLC chromatogram in all systems containing AA compared to the corresponding systems without AA were observed, yet no clear identifiable condensation products between AA with Cy3G and Cy3R were detected.

The stability results of Cy3G and Cy3R indicate that there is a significant increase in the degradation in the presence of AA ($p < 0.05$), with residual concentrations after 146 h of less than 6% for the samples containing AA compared to 77% when AA was not added at pH 4.5 (Figure 4). Figure 4 also reveals that pH has an influence on the degradation rate, as the solution is more acidic the anthocyanin is more stable with and without added AA ($p < 0.05$). The size of the conjugated sugar moiety also has an influence; for both pH values in the presence of AA, Cy3R was more stable than Cy3G, and even more clearly observed at pH 4.5 compared to the system without AA. As the pH value might influence the activity of AA and therefore the degradation of anthocyanins, this point needs further structured in-depth research.

Anthocyanins deterioration is known to follow first-order degradation rate [23–25]. To compare and understand the influence of chemical structure and the presence of AA on the stability of anthocyanins, degradation rate constants (k) were calculated for each explored molecule, with and without the presence of AA at different buffers and temperatures (15 °C, 23 °C and 37 °C). The data are summarized in Table 1. The results present a significant ($p < 0.05$) increase in k values as the storage temperature increased for most systems, with practically no differences between the storage conditions for the lowest studied temperature. The occasionally lacking hypothesized statistically significant differences at the lower temperatures likely originate from the relative stability at the lower temperatures, making statistical verification of differences harder. However, at longer storage times, the presented effects of AA, pH and size of the conjugated sugar are expected to be identifiable also at lower temperatures. The presence of AA significantly accelerated the degradation of anthocyanins stored at 37 °C ($p < 0.05$). At pH = 4.5, without added AA, both Cy3R and Cy3G have no significant differences between the k values in the three explored temperatures, likely due to the relatively high stability at this pH. However, at pH = 6.5, there are significant differences ($p < 0.05$) between the k values at the different temperatures.

Table 1. Degradation rate constant, k (day^{-1}), of Cy3G and Cy3R solutions in the presence of AA at pH = 4.5 and pH = 6.5 stored at 15 °C, 23 °C and 37 °C over time ($n = 2$).

| T (°C) | Cy3G | | | | Cy3R | | | |
| | 4.5 | | 6.5 | | 4.5 | | 6.5 | |
	AA (+)	AA (−)	AA (+)	AA (−)	AA (+)	AA (−)	AA (+)	AA (−)
15	0.036 ± 0.005 [a]	0.057 ± 0.011 [a,c]	0.054 ± 0.010 [a,c]	0.135 ± 0.037 [b,d]	0.043 ±0.006 [a,c]	0.051 ± 0.020 [a,b,c]	0.059 ± 0.003 [c]	0.091 ± 0.041 [d]
23	0.065 ± 0.004 [a]	0.028 ± 0.008 [b]	0.151 ± 0.014 [c,e]	0.127 ± 0.015 [c]	0.095 ± 0.009 [d]	0.027 ± 0.011 [b,e]	0.128 ± 0.012 [e]	0.088 ± 0.008 [a,e]
37	0.659 ± 0.073 [a]	0.024 ± 0.005 [b]	0.922 ± 0.065 [c]	0.528 ± 0.040 [d]	0.469 ± 0.023 [d]	0.057 ± 0.006 [b]	0.774 ± 0.008 [e]	0.402 ± 0.042 [f]

Different letters indicate significant changes at each of the studied temperature.

4. Conclusions

We present a systematic stability study, based on both spectral and HPLC methods, focusing on the effect of cyanidin derivatives structure on the kinetics and outcome of non-enzymatic deterioration. The materials were examined during simulated storage for up to six days in different pH and temperature conditions, with and without AA presence, and the results were analyzed by HPLC-MS and spectral studies. The rapid and high anthocyanins degradation rates, especially when stored without cooling, indicate that it is highly important to identify the mechanisms of non-enzymatic deterioration in the presence of common food formulations, for such health promoting pigments. The anthocyanin chemical structure has an influence on stability and on color, presenting complete lack of stability of the anthocyanidin, resulting in both soluble and insoluble molecules. Further studies are required to fully uncover the mechanism responsible for the higher stability of anthocyanins conjugated to a di-saccharide compared to a mono-saccharide moiety. A decrease in the typical anthocyanin color was observed for all anthocyanins depending on the structure and pH level, yet at the higher pH it was not fully correlated to the original compound degradation as measured by HPLC. Degradation products of Cy were identified and quantified over time by HPLC-MS presenting a complex multi-step degradation reaction. Degradation rate constants were calculated for Cy3G and Cy3R in buffer solutions with and without the addition of AA. When aiming for maximal stabilization of anthocyanins during storage, our results clearly show that the use of AA should be avoided as AA led to a drastic reduction in Cy3G and Cy3R content in both studied pH levels that are representative of mild and low acid products. Complete degradation products profile in buffer solution with and without AA has not been achieved thus far. Future studies should focus not only on the quantification of degradation of the original compounds but also, as was attempted in this work, to identify and quantify the degradation products, in order to better understand the mechanisms allowing improved capabilities for maximizing shelf life. In addition, such research can help in understanding the possible nutritional outcome of anthocyanins deterioration as the degradation products can also have bioactivity.

Author Contributions: Conceptualization, A.S.; Methodology, R.L and Z.O.; Formal analysis, R.L.; Investigation, R.L and Z.O.; Data curation, R.L.; Writing—original draft preparation, R.L.; Writing—review and editing, Z.O and A.S.; Supervision, Z.O and A.S.; Project administration, Z.O.; and funding acquisition, A.S.

Funding: This research was funded by Israel Science Foundation, grant number 686/17.

Conflicts of Interest: The authors declare no conflict of interest.

References

1. Daravingas, G.; Cain, R.F. Thermal degradation of black raspberry anthocyanin pigments in model systems. *J. Food Sci.* **1968**, *33*, 138–142. [CrossRef]
2. Seeram, N.P.; Bourquin, L.D.; Nair, M.G. Degradation products of cyanidin glycosides from tart cherries and their bioactivities. *J. Agric. Food Chem.* **2001**, *49*, 4924–4929. [CrossRef] [PubMed]
3. Timberlake, C.F. Anthocyanins-Occurrence, extraction and chemistry. *Food Chem.* **1980**, *5*, 69–80. [CrossRef]
4. Harborne, J.B. Variation in and functional significance of phenolic conjugation in plants. In *Biochemistry of Plant Phenolics*; Swain, T., Harbone, J.B., Van Sumere, C.F., Eds.; Springer US: Boston, MA, USA, 2012; pp. 457–474.
5. Jackman, R.L.; Yada, R.Y.; Tung, M.A.; Speers, R.A. Anthocyanins as food colorants—A review. *J. Food Biochem.* **1987**, *11*, 201–247. [CrossRef]
6. Joseph, S.V.; Edirisinghe, I.; Burton-Freeman, B.M. Berries: Anti-inflammatory effects in humans. *J. Agric. Food Chem.* **2014**, *62*, 3886–3903. [CrossRef] [PubMed]
7. Folmer, F.; Basavaraju, U.; Jaspars, M.; Hold, G.; El-Omar, E.; Dicato, M.; Diederich, M. Anticancer effects of bioactive berry compounds. *Phytochem. Rev.* **2014**, *13*, 295–322. [CrossRef]
8. Pojer, E.; Mattivi, F.; Johnson, D.; Stockley, C.S. The case for anthocyanin consumption to promote human health: A review. *Compr. Rev. Food Sci. Food Saf.* **2013**, *12*, 483–508. [CrossRef]
9. Castaneda-Ovando, A.; de Pacheco-Hernandez, M.L.; Paez-Hernandez, M.E.; Rodriguez, J.A.; Galan-Vidal, C.A. Chemical studies of anthocyanins: A review. *Food Chem.* **2009**, *113*, 859–871. [CrossRef]
10. Paredes-Lopez, O.; Cervantes-Ceja, M.L.; Vigna-Perez, M.; Hernandez-Perez, T. Berries: Improving human health and healthy aging, and promoting quality life-a review. *Plant Foods Hum. Nutr.* **2010**, *65*, 299–308. [CrossRef] [PubMed]
11. Zafra-Stone, S.; Yasmin, T.; Bagchi, M.; Chatterjee, A.; Vinson, J.A.; Bagchi, D. Berry anthocyanins as novel antioxidants in human health and disease prevention. *Mol. Nutr. Food Res.* **2007**, *51*, 675–683. [CrossRef]

12. Torskangerpoll, K.; Andersen, Ø.M. Colour stability of anthocyanins in aqueous solutions at various pH values. *Food Chem.* **2005**, *89*, 427–440. [CrossRef]
13. Cabrita, L.; Petrov, V.; Pina, F. On the thermal degradation of anthocyanidins: Cyanidin. *RSC Adv.* **2014**, *4*, 18939–18944. [CrossRef]
14. Cabrita, L.; Fossen, T.; Andersen, Ø.M. Colour and stability of the six common anthocyanidin 3-glucosides in aqueous solutions. *Food Chem.* **2000**, *68*, 101–107. [CrossRef]
15. Canuto, G.A.B.; Oliveira, D.R.; Da Conceição, L.S.M.; Farah, J.P.S.; Tavares, M.F.M. Development and validation of a liquid chromatography method for anthocyanins in strawberry (*Fragaria* spp.) and complementary studies on stability, kinetics and antioxidant power. *Food Chem.* **2016**, *192*, 566–574. [CrossRef] [PubMed]
16. Bueno, J.M.; Sáez-Plaza, P.; Ramos-Escudero, F.; Jiménez, A.M.; Fett, R.; Asuero, A.G. Analysis and antioxidant capacity of anthocyanin pigments. Part II: Chemical Structure, color, and intake of anthocyanins. *Crit. Rev. Anal. Chem.* **2012**, *42*, 126–151. [CrossRef]
17. Cortez, R.; Luna-Vital, D.A.; Margulis, D.; Gonzalez de Mejia, E. Natural pigments: Stabilization methods of anthocyanins for food applications. *Compr. Rev. Food Sci. Food Saf.* **2017**, *16*, 180–198. [CrossRef]
18. Rodriguez-Amaya, D.B. Natural food pigments and colorants. *Curr. Opin. Food Sci.* **2016**, *7*, 20–26. [CrossRef]
19. Furtado, P.; Figueiredo, P.; Chaves das Neves, H.; Pina, F. Photochemical and thermal degradation of anthocyanidins. *J. Photochem. Photobiol. A Chem.* **1993**, *75*, 113–118. [CrossRef]
20. Brouillard, R. Chemical structure of anthocyanins. In *Anthocyanins as Food Colors*; Markakis, P., Ed.; Academic Press Inc.: New York, NY, USA, 1982; pp. 1–40.
21. Adams, J.B. Thermal degradation of anthocyanins with particular reference to the 3-glycosides of cyanidin. I. In acidified aqueous solution at 100 °C. *J. Sci. Food Agric.* **1973**, *24*, 747–762. [CrossRef]
22. Markakis, P. Stability of anthocyanins in foods. In *Anthocyanins as Food Colors*; Markakis, P., Ed.; Academic Press Inc.: New York, NY, USA, 1982; pp. 163–180.
23. Sui, X.; Dong, X.; Zhou, W. Combined effect of pH and high temperature on the stability and antioxidant capacity of two anthocyanins in aqueous solution. *Food Chem.* **2014**, *163*, 163–170. [CrossRef]
24. Keith, E.S.; Powers, J.J. Plant pigment-polarographic measurement and thermal decomposition of anthocyanin compounds. *J. Agric. Food Chem.* **1965**, *13*, 577–579. [CrossRef]
25. Marksis, P.; Livingston, G.E.; Fellers, C.R. Quantitative aspects of strawberry pigment degradation. *J. Food Sci.* **1957**, *22*, 117–130. [CrossRef]
26. Iacobucci, G.A.; Sweeny, J.G. The chemistry of anthocyanins, anthocyanidins and related flavylium salts. *Tetrahedron* **1983**, *39*, 3005–3038. [CrossRef]
27. Patras, A.; Brunton, N.P.; O'Donnell, C.; Tiwari, B.K. Effect of thermal processing on anthocyanin stability in foods; mechanisms and kinetics of degradation. *Trends Food Sci. Technol.* **2010**, *21*, 3–11. [CrossRef]
28. Farr, J.E.; Sigurdson, G.T.; Giusti, M.M. Stereochemistry and glycosidic linkages of C3-glycosylations affected the reactivity of cyanidin derivatives. *Food Chem.* **2019**, *278*, 443–451. [CrossRef] [PubMed]
29. Poei-Langston, M.S.; Wrolstad, R.E. Color Degradation in an ascorbic acid-anthocyanin-flavanol model system. *J. Food Sci.* **1981**, *46*, 1218–1236. [CrossRef]
30. Marti, N.; Perez-Vicente, A.; Garcia-Viguera, C. Influence of storage temperature and ascorbic acid addition on pomegranate juice. *J. Sci. Food Agric.* **2002**, *82*, 217–221. [CrossRef]
31. Garcia-Viguera, C.; Bridle, P. Influence of structure on colour stability of anthocyanins and flavylium salts with ascorbic acid. *Food Chem.* **1999**, *64*, 21–26. [CrossRef]
32. Li, J.; Song, H.; Dong, N.; Zhao, G. Degradation kinetics of anthocyanins from purple sweet potato (*Ipomoea batatas* L.) as affected by ascorbic acid. *Food Sci. Biotechnol.* **2014**, *23*, 89–96. [CrossRef]
33. Patras, A.; Brunton, N.P.; Tiwari, B.K.; Butler, F. Stability and degradation kinetics of bioactive compounds and colour in strawberry jam during storage. *Food Bioprocess Technol.* **2011**, *4*, 1245–1252. [CrossRef]
34. Rakic, V.; Skrt, M.; Miljkovic, M.; Kostic, D.; Sokolovic, D.; Poklar-Ulrih, N. Effects of pH on the stability of cyanidin and cyanidin 3-O-β-glucopyranoside in aqueous solution. *Hem. Ind. Chem. Ind.* **2014**, *69*, 511–522. [CrossRef]
35. Nielsen, I.L.F.; Haren, G.R.; Magnussen, E.L.; Dragsted, L.O.; Rasmussen, S.E. Quantification of anthocyanins in commercial black currant juices by simple high-performance liquid chromatography. Investigation of their pH stability and antioxidative potency. *J. Agric. Food Chem.* **2003**, *51*, 5861–5866. [CrossRef] [PubMed]

36. Olivas-Aguirre, F.J.; Rodrigo-García, J.; Martínez-Ruiz, N.D.R.; Cárdenas-Robles, A.I.; Mendoza-Díaz, S.O.; Álvarez-Parrilla, E.; González-Aguilar, G.A.; De La Rosa, L.A.; Ramos-Jiménez, A.; Wall-Medrano, A. Cyanidin-3-O-glucoside: Physical-chemistry, foodomics and health effects. *Molecules* **2016**, *21*, 1264. [CrossRef] [PubMed]

37. Corrales, M.; Butz, P.; Tauscher, B. Anthocyanin condensation reactions under high hydrostatic pressure. *Food Chem.* **2008**, *110*, 627–635. [CrossRef]

38. Kabasakalis, V.; Siopidou, D.; Moshatou, E. Ascorbic acid content of commercial fruit juices and its rate of loss upon storage. *Food Chem.* **2000**, *70*, 325–328. [CrossRef]

39. West, M.E.; Mauer, L.J. Color and chemical stability of a variety of anthocyanins and ascorbic acid in solution and powder forms. *J. Agric. Food Chem.* **2013**, *61*, 4169–4179. [CrossRef] [PubMed]

40. Jurd, L. Some advances in the chemistry of anthocyanin-type plant pigments. In *The Chemistry of Plant Pigments*; Chichester, C.O., Ed.; Academic Press: New York, NY, USA, 1972; pp. 123–191.

 *foods*

Article

Variations in Amino Acid and Protein Profiles in White versus Brown Teff (*Eragrostis Tef*) Seeds, and Effect of Extraction Methods on Protein Yields

Yoseph Asmelash Gebru, Jun Hyun-II, Kim Young-Soo, Kim Myung-Kon and Kim Kwang-Pyo *

Department of Food Science and Technology, College of Agriculture and Life Sciences, Chonbuk National University, Jeonju, Jeollabuk-do 561-756, Korea; holden623@naver.com (Y.A.G.); dnrkrk@hanmail.net (J.H.-I.); ykim@jbnu.ac.kr (K.Y.-S.); kmyuko@jbnu.ac.kr (K.M.-K.)
* Correspondence: kpkim@jbnu.ac.kr; Tel.: +82-63-270-2565

Received: 20 May 2019; Accepted: 7 June 2019; Published: 11 June 2019

Abstract: Data on variations in amino acid compositions and protein profiles among white and brown teff, a grain of growing interest, is either limited or contradicting at the moment. In this study, three white (Addis-W, Mekel-W and Debre-W) and three brown (Addis-B, Mekel-B and Debre-B) teff seed samples were used for whole flour amino acid analysis and protein fractionation with three different methods. White and brown seed types showed different physical changes during protein extraction. Brown teff displayed higher essential amino acid content than white with lysine present in high concentration in both seed types. Extraction with tert-butanol increased prolamin yields in teff compared to ethanol. The major protein fraction in teff was glutelin with white teff containing higher glutelin proportion than brown. Sodium Dodecyl Sulfate Gel Electrophoresis (SDS-PAGE) analysis revealed clear genetic variability between white and brown teff seed types.

Keywords: protein fractionation; white teff; brown teff; amino acid profile; seed storage proteins; essential amino acids

1. Introduction

Teff (*eragrostis tef*) is a small stress tolerant grain (about 0.7% of mass of wheat grain) originally from Ethiopia [1]. Its seed color is either white or very deep reddish brown. In Ethiopia, the second most populous country in Africa, it covers the largest share of area of cereal cultivation with 2.6 million hectares and is a staple food for 80% of the population [2]. Currently, around 200 million people in Ethiopia, Eritrea, Europe and North America consume teff products daily and it is being produced in the USA, Canada, South Africa, Australia and Switzerland [3]. This global demand is a result of its gluten-free nature, high level of essential amino acids (EAA), high mineral content, low glycemic index (GI), high crude fiber content, longer shelf life and slow staling of its bread products compared to wheat, sorghum, rice, barley and maize [4–6].

It is well established that the nutritional and food product qualities of pseudocereals such as amaranth and quinoa as well as common grains are mainly attributed to the physiological functions and food processing characteristics of their seed storage proteins (SSPs) [7], and, more importantly, EAA contents. Amaranth contains 17–19% protein and quinoa and buckwheat were found to be rich in essential amino acids, specially lysine [8,9]. The major proteins in amaranth are albumins (33%) followed by glutelins (30%) and prolamins (3%) [10]. unlike proteins from wheat, rice, maize and barley, pseudocereal proteins do not contain allergens and lysine content of amaranth was found to be twice that of wheat and three times that of maize [11]. On the other hand, quinoa and amaranth flour dough showed preferable rheological properties such as an increase in water absorption and reduction of specific volume compared to wheat flour dough [12], due to higher solubility of their proteins. However, most pseudocereals are usually bushy and slow-growing plants that cannot be harvested as fast as common crops. Teff, a stress tolerant, fast growing grain with similar nutritional qualities, has been recently renewing global interest in pseudocereals.

While it is also rich in protein (12.8–20.9%) and EAA contents (~37%) [13,14], limited and sometimes contradicting data are available on the protein and amino acid profiles of Teff. In terms of SSPs, some literature reported glutelin fraction (45%) as a major protein, followed by albumin (37%) and prolamins (12%) [5], while others claimed albumins to be a major protein fraction in white seed types [15]. More recent papers reported prolamins as a major protein fraction with 40% [4,6]. One study reported that the dominant amino acid in teff is glutamic acid+glutamine (21.8 g/16 g N) followed by alanine (10.1 g/16 g N) and leucine (8.5 g/16 g N) [5]. Besides the few studies on amino acid analysis of crude proteins and distribution of protein fractions, no one has attempted to explore the difference between the white and brown seed types from different regions regarding such profiles.

In this study, we collected six teff samples from 3 different regions in Ethiopia (white and brown samples from each region) and analyzed their amino acid compositions, which allowed, for the first time, comparative nutritional qualities among teff seed types. In addition, we compared three different protein extraction methods to explain the previously contradicting results on major protein fraction in Teff. Further analysis of differently fractionated proteins on SDS-PAGE showed significant difference in protein banding patterns suggesting possible quality or functional differences among teff seed types.

2. Materials and Methods

2.1. Preparation of Plant Materials

Teff seed samples harvested during 2017 harvest season were purchased from local markets in Addis Ababa (Central Ethiopia; white called Addis-W and brown called Addis-B), Mekelle (Northern Ethiopia; Mekel-W and Mekel-B) and Debremarkos (Western Ethiopia; Debre-W and Debre-B), and brought to South Korea with proper packaging. Whole grains were finely ground using a roll mill (Single type stainless roller, Shinpoong Eng. Ltd., Gwangju, Gyeonggi, Korea) with 0 mm gap between the rollers and 4 smashes. The flour was again sieved through a 1mm mesh sieve and stored at 4 °C until use.

2.2. Chemicals and Reagents

All reagents and buffers were of analytical grade and purchased from Sigma-Aldrich (St. Louis, MO, USA) unless indicated otherwise.

2.3. Amino Acid Analysis

Amino acid compositions of flour samples were determined according to a modified method of Spackman et al. [16]. For hydrolysis, 0.5 g of powdered seed samples were dissolved in 6 N HCl solution and heated for 24 h at 110 °C. The solution was evaporated under reduced pressure using a rotary evaporator (EYELA Rotary vacuum evaporator N-N SERIES, Tokyo Ridadidai Co. Ltd, Tokyo, Japan) and the residual solid was dissolved in sample dilution buffer (pH 2.2). The solution was filtered through a 0.45 µm membrane (ADVANTEC Toyo Roshi Kaisha, Japan). Amino acid analysis was performed using a Sykam (Sykam Co., Darmstadt, Germany) S7130 amino acid reagent organizer, S5200 sample injector and S2100 solvent delivery system with a cation separation column LCA K06/NA (4.6 mm × 250 mm). The flow rates of mobile phase and ninhydrin were 0.45 mL/min and 0.4 mL/min, respectively. Due to the acid hydrolysis, glutamine and asparagine were determined with glutamic acid and aspartic acid, respectively, and tryptophan was degraded and not detected.

2.4. Fractionation of Teff Proteins

Three types of methods were used to sequentially extract four fractions of proteins (albumins, globulins, prolamins and glutelins) from teff seed flour. Method 1 was conducted based on a modified protein fractionation method described by Ayoni et al. [17]. Briefly, albumins were extracted with deionized water on flour to solvent ratio of 1:10 (*w/v*) three times by shaking for 1 h each time and a fourth time for 30 minutes. Each extract was centrifuged at 6000× *g* for 10 min at 4 °C to obtain a clear supernatant, and all supernatants were combined as albumin fraction. The residue after albumin extraction was used to extract globulins using 1.25 M NaCl in a similar procedure for albumins. After washing the residue from the globulin extraction 2 times for 1 h with deionized water to remove salt, prolamins were extracted from the residue using 70% ethanol. Again, after washing with deionized water the same way as above to remove the alcohol, glutelins were extracted with 0.075 M NaOH in similar steps as above.

Method 2 was done according to a modified version of a method described by Wallace et al. [18]. Albumins and globulins were extracted exactly the same as in method 1. Prolamins were extracted using 70% ethanol containing 5% β-merkaptoethanol (βME) at RT (23 °C) 3 times for 1 h and a fourth extraction done overnight. Glutelins were extracted with 0.075 M NaOH containing 5% βME and 1% SDS the same way as prolamins in this method.

Method 3 was performed using a modified method by Tylor et al. [19]. Albumins and globulins were extracted exactly the same was as in Method 1 and 2. Prolamins were extracted using 60% tert-butanol containing 0.05% DTT at 23 °C and the fourth time extraction was done overnight. Similarly glutelins were extracted with 0.075M NaOH containing 0.05% DTT at 23 °C.

Each supernatant was filtered through a 0.45 μm membrane (ADVANTEC Toyo Roshi Kaisha, Japan) to remove insoluble particles and then dialyzed against distilled water with a 3.5 kDa MW cut of dialysis membrane (Spectrum Laboratories, Rancho Dominguez, CA, USA) with distilled water for 24 hours and 4 changes of water. Protein samples were freeze-dried (MCFD8512; Ilshinbiobase Co., Ltd., Gyeonggi, Korea) and stored at −20 °C until used for SDS-PAGE analysis.

2.5. Determination of Protein Concentrations

Protein concentrations were determined as described by Bradford et al. using bovine serum albumin (BSA) as a standard [20]. Protein samples were diluted to fit the absorption range of the standard concentrations to a final volume of 100 μL and mixed with 900 μL of Bradford reagent (Sigma, B6916). After vigorous mixing and incubating at room temperature for 10 minutes, absorbance was measured at 595 nm using UV-visible spectrophotometer (Biochrom–Libra S22, Cambridge, UK).

2.6. Sodium Dodecyl Sulfate Gel Electrophoresis (SDS-PAGE)

To determine the molecular weight (MW) distribution of each protein fraction, SDS-PAGE was conducted as described by Laemmli [21]. Extracted samples were filtered through a 0.45 μm membrane, dialyzed against distilled water with a 3.5 kDa MW cut of dialysis membrane and freeze dried. Premixed polyacrylamide gel kit was purchased from Bio-Rad (TGX™ FastCast™ Acrylamide Solutions, Bio-Rad Corporations, Hercules, CA, USA). Protein fractions were run in a 5% stacking gel (*w/v* polyacrylamide) and 15% separating gel. 17 μg of protein was loaded per well and run on AE-6531 PAGERUN mini slab electrophoresis system (ATTO Corporation, Tokyo, Japan) for 2 hours at 40 mA. Protein bands were then visualized after staining the gel with coomassie brilliant blue R-250 and MW of prominent bands were estimated by comparing to the Precision Plus Protein Standard (Bio-Rad Laboratories, Hercules, CA, USA).

2.7. Statistical Analysis

All experiments were conducted in triplicate and the results were represented as mean ± standard deviation (SD). One way ANOVA analysis of the mean and variance of the extraction yields was done using Graph pad prism 5 (GraphPad Software, San Diego, CA, USA) to compare the efficiency of methods and distribution of protein fractions and amino acids in white versus brown teff. Differences with $p < 0.05$ were considered significant.

3. Results

3.1. Physical Characteristics of Teff Seeds

All six teff seed samples were ground and used for three different protein fractionation methods, during which color changes in residues and supernatants were monitored. The pictorial data from Mekel-W and Mekel-B during Method 3 extraction procedures were selected as representatives because the same results were observed for all samples (See below). The two seed types have indubitably different colors with linen white and brown colors, respectively (Figure 1a,b). However, the difference in color became less pure when finely ground (Figure 1c,d).

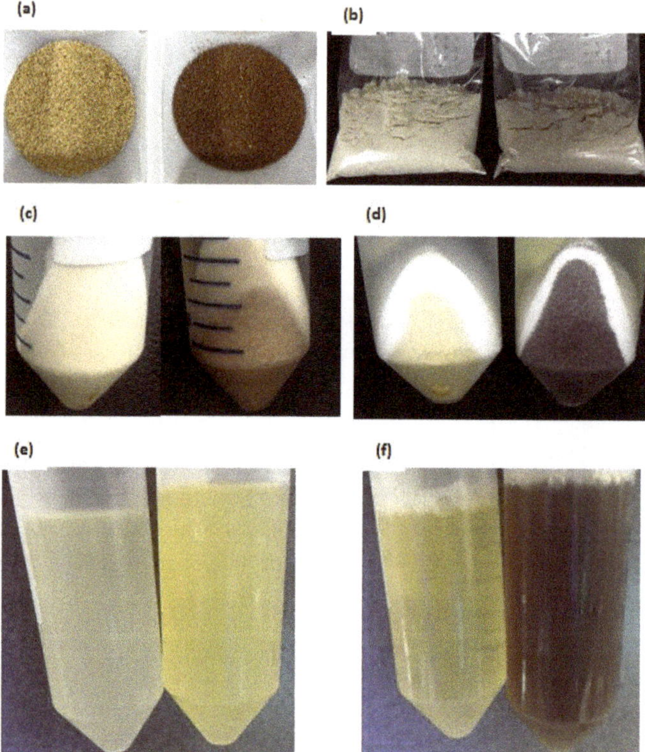

Figure 1. Physical appearances of Mekel-W and Mekel-B during sample preparation. (**a**) Seeds (left to right; Mekel-W, Mekel-B), (**b**) Teff seed flour (left to right; Mekel-W, Mekel-B), (**c**) Residue before ethanol extraction (left to right; Mekel-W, Mekel-B), (**d**) Residue after ethanol extraction (left to right; Mekel-W, Mekel-B), (**e**) tert-butanol extract supernatants (left to right; Mekel-W, Mekel-B), (**f**) NaOH extract supernatants (left to right; Mekel-W, Mekel-B).

No change in color of residue was observed after albumins (water extraction) or globulins (NaCl extraction) were removed while white teff residue became more elastic and viscous than brown. After alcohol extraction for prolamin fraction, however, the color of the residue of brown teff flour changed from reddish brown to a dark umber color while the white type remained same color (Figure 1e,f). The supernatants of water and salt extractions of both seed types were colorless, while the alcohol extract supernatants of white and brown teff flours were of light and yellowish color, respectively (Figure 1g). After alkaline (0.075M NaOH) extraction, the color of the supernatant was a dark brown or amber color (Figure 1h).

3.2. Amino Acid Compositions

Six whole seed flour samples (Addis-W, Addis-B, Mekel-W, Mekel-B, Debre-W and Debre-B) were used for total amino acid analysis and amount of 17 amino acids were determined (Table 1). A representative HPLC chromatogram for amino acid profile of teff seed flour is also shown in Figure 2. The total amino acid contents of all brown teff samples were higher than that of white. No clear distinction in total amino acid contents among the samples of same seed color was observed. The contents of individual amino acids were highly correlated to their total amino acid contents and generally higher in brown teff than that of white teff. With regard to the ratio of each of the amino acids, glutamic acid and glutamine together were the dominant amino acids in all samples, accounting for an average of 26.35% of total amino acid contents followed by aspartic acid and asparagine with average ratio of 9.16%.

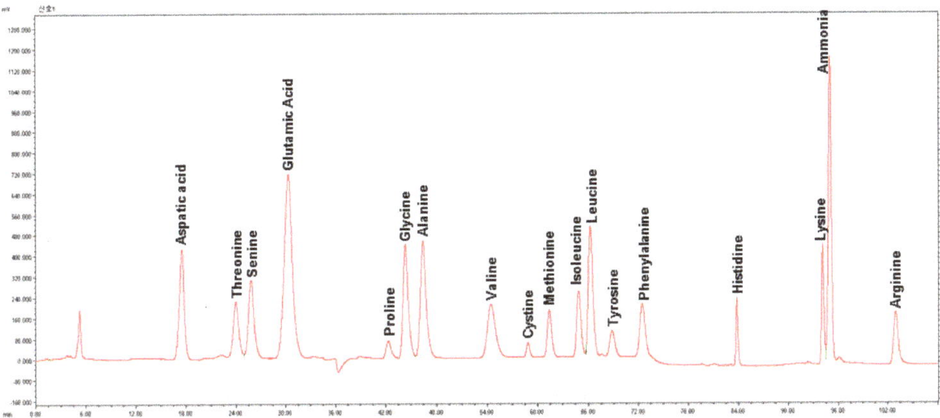

Figure 2. Representative chromatogram showing amino acid profile of teff seed flour (Addis-W). Y-axis is intensity (mAu); X-axis is retention time (min).

The ratio of individual amino acids within each sample showed no significant difference among all 6 samples with the exception of histidine which showed higher proportion in Debre-B (6.96%) than others, and lysine which was higher in Debre-W (9.37%) than others. In addition, arginine was found to be in higher proportion in white teff than in brown, unlike the other amino acids in which their ratios were consistently similar in all samples of both seed types (Table 1).

The total essential amino acid (EAA) contents of brown teff were higher than white teff. Generally, there was no significant difference in EAA content among the samples of same seed color and ratio among all samples. An exception was Debre-B which showed a notably higher total EAA content (104.50 mg/g flour dry weight) than the other brown teff samples (90.34 and 84.45 for Addis-B and Mekel-B respectively). The ratio of EAA in this particular sample was 46.60% which is far from the average (40.40 and 40.90 for white and brown, respectively).

Among the individual EAAs, lysine (average of 12.82 mg/g of flour dry weight) was at the highest concentration in white teff, followed by leucine (11.74), isoleucine (9.34), phenylalanine (8.52), threonine (7.05), valine (6.02), methionine (4.52) and histidine (3.02). In brown teff the highest amount was leucine (17.51), followed by lysine (15.12), isoleucine (14.06), phenylalanine (12.52), threonine (10.40), valine (8.73), histidine (8.06) and methionine (6.70).

Table 1. Amino acid composition (mg/g flour dry weight) and ratio of individual amino acid (%) of 6 teff flour samples.

Amino Acid		Samples											
		Addis-W	Ratio (%)	Addis-B	Ratio (%)	Mekel-W	Ratio (%)	Mekel-B	Ratio (%)	Debre-W	Ratio (%)	Debre-B	Ratio (%)
Essential Amino Acids	Thr	6.73 ± 0.35	4.35	10.76 ± 0.66	4.48	7.53 ± 0.15	4.68	10.45 ± 0.07	4.78	6.90 ± 0.14	4.52	10.00 ± 0.00	4.46
	Val	5.60 ± 0.20	3.62	8.63 ± 0.32	3.59	5.66 ± 0.32	3.52	8.35 ± 0.21	3.82	6.80 ± 0.28	4.46	9.20 ± 0.84	4.1
	Met	4.10 ± 0.17	2.65	7.16 ± 0.32	2.98	4.66 ± 0.49	2.9	6.15 ± 0.35	2.81	4.80 ± 0	3.15	6.80 ± 0.28	3.03
	Ile	9.23 ± 0.05	5.96	14.13 ± 0.11	5.88	9.20 ± 0.45	5.72	13.85 ± 0.63	6.33	9.60 ± 0.00	6.29	14.20 ± 0.28	6.33
	Leu	11.40 ± 0.26	7.36	17.53 ± 0.28	7.29	11.73 ± 0.64	7.29	16.70 ± 0.14	7.64	12.10 ± 0.42	7.93	18.30 ± 0.42	8.16
	Phe	8.23 ± 0.58	5.31	12.60 ± 0.36	5.24	8.43 ± 0.3	5.24	11.35 ± 0.49	5.19	8.90 ± 0.70	5.83	13.60 ± 0.84	6.07
	His	2.96 ± 0.05	1.91	4.43 ± 0.05	1.84	3.13 ± 0.11	1.94	4.15 ± 0.21	1.9	3.10 ± 0.14	2.03	15.60 ± 1.69	6.96
	Lys	12.00 ± 0.17	7.75	15.10 ± 0.40	6.28	12.16 ± 0.6	7.56	13.45 ± 0.49	6.15	14.30 ± 0.70	9.37	16.80 ± 0.00	7.49
	Subtotal	60.25	38.91	90.34	37.58	62.5	38.85	84.45	38.62	66.5	43.58	104.5	46.6
Other amino acids	Asp	15.93 ± 0.55	10.29	19.23 ± 0.86	8	15.06 ± 0.75	9.36	18.30 ± 0.98	8.37	15.60 ± 0.84	10.22	19.60 ± 0.00	8.74
	Ser	7.16 ± 0.55	4.62	11.06 ± 0.66	4.6	7.86 ± 0.25	4.88	10.90 ± 0.14	4.99	7.10 ± 0.42	4.65	11.20 ± 0.00	5
	Glu	38.90 ± 0.34	25.12	75.86 ± 8.25	31.56	43.86 ± 1.78	27.25	63.35 ± 9.26	28.97	33.50 ± 0.98	21.95	52.10 ± 2.68	23.24
	Pro	1.86 ± 0.11	1.2	2.96 ± 0.32	1.23	1.96 ± 0.32	1.22	2.95 ± 0.07	1.35	2.10 ± 0.42	1.38	3.40 ± 0.00	1.52
	Gly	5.76 ± 0.15	3.72	7.50 ± 0.26	3.12	6.00 ± 0.40	3.73	6.90 ± 0.28	3.16	5.90 ± 0.42	3.87	7.60 ± 0.28	3.39
	Ala	7.06 ± 0.15	4.56	10.40 ± 0.26	4.33	7.50 ± 0.51	4.66	12.15 ± 2.75	5.56	9.00 ± 0.28	5.9	10.40 ± 0.00	4.64
	Cys	0.63 ± 0.05	0.41	1.13 ± 0.05	0.47	0.63 ± 0.05	0.39	0.85 ± 0.07	0.39	0.4 ± 0.00	0.26	ND	ND
	Tyr	3.36 ± 0.05	2.17	5.46 ± 0.05	2.27	3.43 ± 0.15	2.13	5.05 ± 0.07	2.31	ND	ND	3.60 ± 0.00	1.61
	Arg	13.96 ± 1.22	9.01	16.43 ± 0.32	6.84	12.13 ± 0.75	7.54	13.75 ± 0.49	6.29	12.5 ± 0.98	8.19	11.80 ± 1.41	5.26
	Subtotal	94.62	61.1	150.03	62.42	98.43	61.16	134.2	61.39	86.1	56.42	119.7	53.4
Total		154.87 ± 1.01	100	240.37 ± 4.17	100	160.93 ± 7.80	100	218.65 ± 8.27	100	152.6 ± 4.24	100	224.2 ± 14.56	100

Values are mean ± standard deviation of triplicate. ($p < 0.05$). Thr = Threonine; Val = Valine; Met = Methionine; Ile = Isoleucine; Leu = Leucine; Phe = Phenylalanine; His = Histidine; Lys = Lysine; Asp = Aspartic acid; Ser= Serine; Glu = Glutamic acid; Pro = Proline; Gly = Glycine; Ala = Alanine; Cys = Cysteine; Tyr = Tyrosine; Arg = Arginine.

3.3. Effect of Extraction Methods on the Yield of TSSP

The yields of Teff seed storage protein fraction (TSSPF) prepared by three methods with different extraction solvent compositions were compared and are shown in Table 2.

Table 2. Effects of different extraction methods on yield of Teff seed storage protein fractions (TSSPF).

Sample	TSSPF	Method 1		Method 2		Method 3	
		Yield (g/100g flour)	Ratio (%)	Yield (g/100g flour)	Ratio (%)	Yield (g/100g flour)	Ratio (%)
	Albumin	2.13 ± 0.22	31.88	2.13 ± 0.16	29.7	2.06 ± 0.09	23.01
	Globulin	0.38 ± 0.01	5.71	0.38 ± 0.01	5.31	0.36 ± 0.05	3.97
Addis-W	Prolamin	0.10 ± 0.03 [a]	1.46	0.43 ± 0.01 [b]	5.96	1.62 ± 0.18 [c]	18.09
	Glutelin	4.07 ± 0.30 [a]	60.94	4.23 ± 0.25 [a]	59.04	4.93 ± 0.19 [b]	54.92
	Total	6.69 ± 0.55 [a]	100	7.16 ± 0.41 [a]	100	8.97 ± 0.27 [b]	100
	Albumin	2.43 ± 0.29	38.37	2.50 ± 0.24	33.97	2.53 ± 0.15	26.98
	Globulin	0.52 ± 0.12	8.28	0.55 ± 0.08	7.44	0.38 ± 0.29	4.08
Addis-B	Prolamin	0.19 ± 0.03 [a]	3.02	0.94 ± 0.02 [b]	12.77	2.37 ± 0.32 [c]	25.29
	Glutelin	3.19 ± 0.18 [a]	50.33	3.37 ± 0.22 [a]	45.81	4.09 ± 0.27 [b]	43.64
	Total	6.34 ± 0.43 [a]	100	7.35 ± 0.38 [b]	100	9.37 ± 0.39 [c]	100
	Albumin	1.88 ± 0.14	41.81	2.01 ± 0.13	38.13	2.10 ± 0.16	31.37
	Globulin	0.37 ± 0.01	8.26	0.26 ± 0.04	5.01	0.30 ± 0.01	4.46
Mekel-W	Prolamin	0.19 ± 0.01 [a]	4.12	0.68 ± 0.05 [b]	12.98	1.75 ± 0.11 [c]	26.22
	Glutelin	2.06 ± 0.08 [a]	45.81	2.31 ± 0.10 [bc]	43.88	2.54 ± 0.10 [c]	37.95
	Total	4.50 ± 0.23 [a]	100	5.26 ± 0.19 [b]	100	6.69 ± 0.36 [c]	100
	Albumin	1.53 ± 0.21	38.84	1.26 ± 0.17	25.36	1.71 ± 0.23	26.07
	Globulin	0.17 ± 0.02	4.27	0.14 ± 0.02	2.88	0.13 ± 0.02	2.06
Mekel-B	Prolamin	0.24 ± 0.01 [a]	6	1.23 ± 0.10 [b]	24.73	2.24 ± 0.05 [c]	34.16
	Glutelin	2.01 ± 0.08 [a]	50.89	2.33 ± 0.07 [bc]	47.03	2.47 ± 0.10 [c]	37.71
	Total	3.94 ± 0.16 [a]	100	4.96 ± 0.32 [b]	100	6.55 ± 0.21 [c]	100
	Albumin	1.82 ± 0.13	27.57	2.36 ± 0.14	30.8	2.42 ± 0.19	29.53
	Globulin	0.29 ± 0.04	4.36	0.39 ± 0.01	5.13	0.38 ± 0.07	4.62
Debre-W	Prolamin	0.23 ± 0.02 [a]	3.5	0.46 ± 0.01 [b]	5.96	1.55 ± 0.28 [c]	18.87
	Glutelin	4.27 ± 0.20	64.57	4.44 ± 0.14	58.11	3.85 ± 0.18	46.98
	Total	6.61 ± 0.30 [a]	100	7.65 ± 0.25 [ab]	100	8.20 ± 0.62 [b]	100
	Albumin	2.97 ± 0.27	42.14	2.82 ± 0.21	35.78	2.65 ± 0.16	30.29
	Globulin	0.50 ± 0.10	7.03	0.61 ± 0.08	7.75	0.44 ± 0.07	5.05
Debre-B	Prolamin	0.38 ± 0.05 [a]	5.39	1.02 ± 0.03 [b]	12.98	2.51 ± 0.22 [c]	28.71
	Glutelin	3.20 ± 0.23	45.43	3.42 ± 0.19	43.48	3.15 ± 0.18	35.95
	Total	7.05 ± 0.53 [a]	100	7.87 ± 0.35 [ab]	100	8.75 ± 0.14 [b]	100

Values are mean ± SD of triplicate. Values with different superscript alphabets within the same row under protein yields are significantly different ($p < 0.05$).

In Method 1, among Addis Ababa and Debremarkos samples, amounts of total TSSP yields (6.34–7.05 g/100 g flour) and each TSSPF (1.82–2.97 for Albumins, 0.29–0.52 for globulins, 0.1–0.38 for prolamins, 3.19–4.27 g/100 g for glutelins) were very similar regardless of seed colors. On the other hand, significantly lower amounts of total TSSP (4.50 and 3.94 for Mekel-W and Mekel-B, respectively) were extracted from Mekelle samples. In all the samples, glutelin was the most prominent fraction followed by albumins and globulin (or prolamins). In this method, there was no difference in the total protein yield between white and brown seed types. However, the average ratio of glutelins proportion in white teff (57.11%) was higher than that of in brown teff (48.88). There was no significant difference in other fractions between white and brown.

In Method 2, higher amounts of total TSSP from all the samples (7.16–7.87 g/100 g for Addis Ababa and Debremarkos samples which is significantly higher than 4.96–5.26 g/100 g for Mekelle samples) were extracted compared to Method 1. This was mainly due to the increase in prolamin and glutelin yields. The most notable difference compared to Method 1 was a more profound increase in prolamin fraction, leading to significant increase especially for Mekel-B sample (0.24–1.23 g/100 g) in fractional yield for prolamin. Nevertheless, the most prominent fraction is Glutelin followed by albumins, prolamins and globulins. Similarly, as in Method 1, there was no difference in the total protein yield between white and brown teff. The average ratio of glutelins in white teff (53.68%) was higher than that of in brown teff (45.44%). On the other hand, the average prolamin content (16.83%) in brown teff was higher than that of white (8.3%). Albumin and globulin remained similar.

In Method 3, most efficient extractions of total TSSP (8.20–9.37 g/100 g for Addis Ababa and Debremarkos samples which is significantly higher than 6.55–6.69 for Mekelle samples) from all the samples were observed compared to the others. Notably, in all the samples, significantly higher ($p < 0.0001$) increases in prolamin fraction yields were observed, even making it to a change in order of magnitude (Mekel-B; for Method 1 vs. Method 3). Among all samples, Mekel-B had significantly higher ($p < 0.05$) prolamin content (2.24 g/100 g or 34.16%) than other samples. Glutelins were major proteins followed by albumins, prolamins and globulins except in Mekelle-B where prolamins held the second place. As in Methods 1 & 2, there was no difference in the total protein yield between white and brown teff. The average ratio of glutelins in white (46.62%) was also higher than that of the brown teff (39.1%). The average prolamin content in brown teff (29.39%) was higher than that of white (21.06%). Albumin and globulin remained similar among the two seed types.

3.4. Molecular Weight Distribution of Teff Proteins

Protein fractions extracted with Method 3 are shown for SDS-PAGE analysis as, in general, very similar protein patterns were observed for the same sample extracted with different methods (Figure 3). One exception was prolamin fraction which showed notable difference between Method 3 and the other two methods (See below).

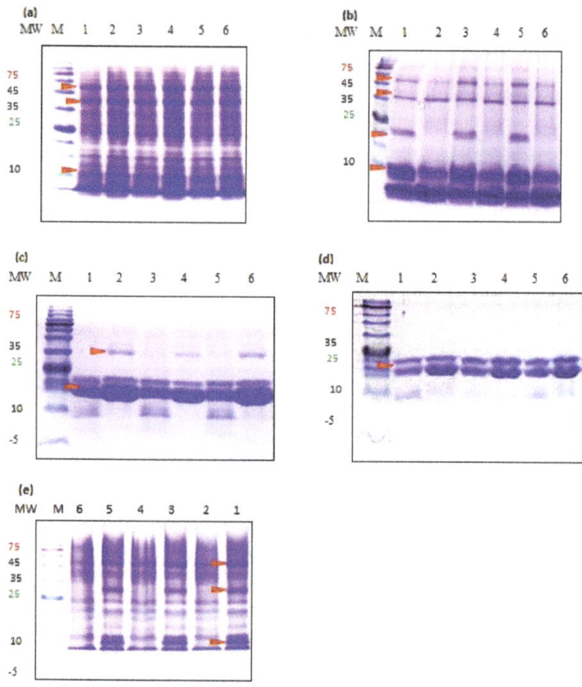

Figure 3. Sodium Dodecyl Sulfate Gel Electrophoresis band patterns of teff protein fractions resolved on a 15% gel under reducing conditions. (**a**) Albumins, (**b**) Globulins, (**c**) Prolamins Method 2, (**d**) Prolamins Method 3, (**e**) Glutelins. MW, molecular weight (kilodaltons); M, marker; Lane 1, Addis-W; Lane 2, Addis-B; Lane 3, Mekel-W; Lane 4, Mekel-B; Lane 5, Debre-W; Lane 6, Debre-B.

Albumin fraction was resolved to show no variation between white and brown types of different samples (Figure 3a). The highest number of polypeptides was observed among other fractions with MW ranging from 10–100 kDa. The three most prominent bands were approximately at 10, 53 and 67 kDa in size.

Proteins in globulin fraction were characterized by the low MW proteins being dominant. Regardless of different sample regions, the same banding patterns were observed among the same seed color types, but a clear difference was found between white and brown types (Figure 3b). In white seed type, the fraction included 5 subunits approximately at 6, 9, 22, 37.5 and 52.5 kDa. In the case of brown type, all bands were detected at same locations with similar intensity as in white type with the exception of the 22 kDa subunit which was completely absent.

Proteins in Prolamin fraction were characterized by a high expression level of low MW proteins. In Methods 1 and 2, regardless of different samples regions, four bands approximately at 9.5, 11, 19.5 and 23 kDa for white type and three bands at 19.5, 23 and 35 kDa for brown type were detected (Figure 3c). In Method 3, only the 19.5 and 23 kDa bands were detected in similar patterns in both seed types (Figure 3d). The 23 kDa band which formed a very intense band in both seed types had significantly higher intensity in brown teff than in white.

Proteins in Glutelin fractions were also diverse, as in albumin fraction, with molecular sizes ranging from approximately 7 to 100 kDa (Figure 3d). Again, similar banding patterns were observed among the samples of different regions, but clear difference was found between seed color types. Most notable difference was the presence of 9 and 30 kDa bands in brown type which are absent in white.

4. Discussions

4.1. Physical Characteristics

A less visible difference is observed in the flour colors of white and brown teff (Figure 1c,d) compared to the difference in their respective seed colors (Figure 1a,b), indicating that the pigmenting compounds of brown teff are mainly accumulated on the grain pericarp. Previously, it was shown that the endosperm of both seed types have similar composition which is predominantly composed of starch, while the pericarp of brown teff contains various compounds that give the seed its distinct color [22].

We also found the changes in the colors of the residue to deep umber (Figure 1e,f) and of the supernatant to light yellowish color of brown teff after alcohol extraction (Figure 1g). A recent study found that while an unusually high level of flavones are present in both white and brown teff, tannins are present as procyanidin (condensed tannins) form only in the brown seed [22]. The fact found in our study that the color of the supernatant turned dark umber (Figure 1h) in the NaOH extract may also suggest that the insoluble tannins are partially soluble in alkaline solution, which is similar to previous studies [23,24]. Further explorations on the ethanol extract and the procyanidins in brown teff may reveal a variety of bioactive compounds accumulated on teff bran and discover more compounds responsible for the color.

4.2. Amino Acid Compositions

It is well accepted that the dietary value of SSPs is measured depending on the amino acid composition [25]. In this study, we found that the amino acid compositions of teff were very similar among the same seed color types (white or brown) from different regions while total amino acid content of brown teff was found to be significantly higher ($p < 0.05$) than the white one. Altogether, the total amino acid content of teff (227.74 and 154.87 mg/g flour for brown and white teff respectively) is higher than amaranth (140.1) and quinoa (113.9) [26] which can be a good reason to attract interest on teff research together with these pseudocereals.

Despite the importance of total protein contents, essential amino acid content is more crucial characteristic when assessing the dietary quality of grains. In this study, the total amount of essential amino acids in brown teff was considerably higher than that of white seed type, while the overall ratio of essential amino acids relative to the remaining amino acids was similar in both white and brown (40.4% and 40.88% respectively). Compared to other grains such as wheat (41.5) [27], barely (21.8), maize (25.6) [28], rice 37.8 mg/g of flour [29] etc., Teff seed flour is relatively rich in EAA (62.5–104.5 mg/g of flour) with a well-balanced concentration (Table 1). EAA content and balance has very important value in common cereals which are sometimes the sole source of nitrogen in the developing world.

Lysine, an essential amino acid which exists in a limited amount in other cereals was observed to be in higher concentration (~12–16 mg/g of flour) in teff. This is much higher content compared to the pseudocereals amaranth (7) and quinoa (8.3 mg/g flour) [26]. The fact that lysine exists in limited content in these common cereals which are the staple foods for most of world population is a big challenge in protein diet [30]. Therefore such high lysine content in teff can be considered as a key factor in introducing new grains with high nutritional value.

As shown in the protein fractionation section, the total yield of protein extracts did not show significant difference between white and brown seeds (in spite of significant difference between different region samples of the same color, See below) (Table 2). However the concentration of total amino acids was found to be significantly different between white and brown types, brown teff being higher than white one. This shows that the content of amino acids are independent of the total amount of extracted protein fractions in the seed flour, which was also reported for the pseudocereal quinoa [31]. The possible reason for such miscorrelation in teff is suggested to be the interaction between the insoluble tannins and proteins in brown teff that may have caused insolubility during extraction.

Overall, the higher concentration of total and essential amino acids in brown teff can be considered as its nutritional advantage over the white one. To the best of our knowledge, this is the first study to present an organized comparison of amino acid profiles between white and brown teff seeds and superiority of brown teff.

4.3. Effect of Extraction Methods on the Yield of TSSP

SSPs are not only nutritionally important but also influence the utilization of grains in food processing. Their characteristics are more important in grains such as wheat and teff which are consumed after processing into various kinds of foods. This called for exploring the less characterized TSSPs to review the existing contradicting data on their fractions.

In the case of prolamins, Method 3 resulted in significantly higher ($p < 0.001$) yield followed by Method 2 and then Method 1 (Table 2). This showed that tert-butanol was much more effective than ethanol in extracting alcohol soluble proteins from cereals. Tert-butanol has been designated as a natural gift for protein isolation for its advantage in stabilizing protein structures during extraction. This might be attributed to its larger size that can hinder it from accessing the interior of the protein which stabilizes the protein instead of denaturing. It also inhibits enzyme activities and protein-protein interactions minimizing formation of artifacts and resulting in higher extraction yield [31,32]. Based on this, it can be assumed that the increase in prolamin yield is due to an overall improved stability and less proteolysis of the protein fraction.

Method 2 also significantly ($p < 0.05$) increased prolamin yield compared with Method 1 which is attributed to the reducing agent βME (Table 2). βME is a strong reducing agents that cleave disulfide crosslinks in proteins and also inhibit oxidation of free sulfhydryl residues. The increased prolamin yield in Method 2 (average 0.79 g/100 g flour) from Method 1 (0.09) indicate that teff prolamines are prone to oxidative damage and can be effectively protected by reducing agents during extraction. This protein fraction has been reported to form disulfide bonds during extraction and reducing agents effectively enhanced their extraction yield by preventing the formation of disulfide bonds [33]. Here, it can be assumed that the increase in prolamin yield is due to an increase in overall solubility and a decrease in the oxidation damage of the fraction.

In the case of glutelin fraction, there was only a slight difference in yield between Method 1 and Method 3 but not between Method 1 and 2 (Table 2). In grains such as maize, glutelin has shown to be highly insoluble in the most potent protein dissociating solvents [34]. However, with reduction of disulfide bonds by βME in the extraction solution, it was possible to increase its yield [35]. In this study, glutelin fraction was extracted using 0.075 M NaOH with or without the reducing agents βME and DTT. Even though DTT showed a relatively better effect than βME similar to the previous study [36], there was no significant difference compared to Method 1 (no reducing agent). This indicates that the glutelin fraction in a gluten free teff seed is relatively readily soluble compared to the glutelin in maize (a high gluten grain). It has been well established that glutelin is the major component of gluten and the gluten level in cereals is linearly related to the percentage of disulfide bonds in the glutelin fraction [37]. Therefore, the gluten free nature of teff and the no effect observed with or without reducing agents during extraction could be attributed to a less percentage of disulfide crosslinks in teff glutelin proteins.

In method 1 the ratio was glutelins > albumins > globulins > prolamins in which globulins and prolamins comprised only a small portion. The distribution order in method 2 and 3 was glutelins > albumins > prolamins > globulins. However, the relative proportion of prolamins and albumins in method 3 was almost similar while their variation gap was wider in method 2. This is partially in agreement with [5,38] who reported glutelins as major proteins (45%) followed by albumins (37%) and prolamins (12%), yet it is in contrast with a result reported by Abdul-Rasaq et al. [4] where prolamin was the major fraction accounting for 40% of the total protein in teff and Zhang et al. [6] which suggested prolamin was the major protein after analyzing the amino acid composition of teff. It is common to obtain different results of protein proportions with different extraction methods and samples from growth environments. Even though we used the same solvent as Abdul-Rasaq et al. [4], the different result might be due to different samples. In one of our samples (Mekel-B) prolamin ratio (34.16%) was almost similar with glutelin (37.71%).

In Method 3, white teff was found to have significantly higher glutelin content (46.62%) than brown (39.1%) on average. Previous studies have confirmed that glutelin together with prolamin plays an important role in bread making characteristics in wheat and other grains [39]. We observed that the rheology of the residue of white teff before glutelin extraction was much more viscous and elastic compared to brown. Therefore, the higher proportion of glutelins in white teff may have a big influence on functional properties including baking performance and dough rheology.

On the other hand, the average prolamin content (29.39%) of brown teff in Method 3 was higher than that of white teff (21.06%). It has been proven that prolamins play a great role in aggregation of protein bodies (PBs) in teff and maize endosperms as examined by transmission electron microscopy (TEM) [6]. Therefore, the higher expression of prolamins in brown teff may also influence the functional properties in food processing. Similarly, prolamins with their distinctive amino acid compositions can alter the overall proportion of essential amino acids in brown teff thereby bringing nutritional difference. In this study we also proved that brown teff contains significantly higher essential amino acids content than white teff (Table 1). To have a deeper understanding of the difference in physical and nutritional properties between white and brown varieties, a more detailed study on teff prolamins is required.

While the total protein yields of Addis Ababa and Debremarkos samples were somehow similar (Table 1), the samples from Mekelle showed significantly lower total protein yield. It is common to observe variations in protein content across locations while amino acids are stable [40]. Therefore, the difference might have been caused by different environmental conditions during plant development.

4.4. SDS-PAGE

SDS PAGE analysis was conducted to examine any variation of polypeptide patterns. Despite different total amounts of protein fractions among the samples from different regions (Table 2), no quantitative or qualitative variation was observed among the same seed types from different regions. However, clearly visible differences in band numbers and band quantities were observed between white and brown seed types, except for the albumin fraction which showed similar patterns. Our prolamin patterns are in partial agreement with Abdul Rasak et al who detected the 19.5 and 23 kDa subunits for both white and brown teff at similar MW [4]. Zhang et al. also reported similar results for brown teff with two most predominant bands detected at 19 & 22 kDa [6]. We could not find previous SDS PAGE patterns of TSSPFs in the literature except for prolamins.

Our study discovered significant variation in protein profiles between white and brown teff based on their SDS PAGE patterns of globulin, prolamin and glutelin fractions. The SDS PAGE pattern variation between white and brown teff can be used as a tool to study genetic diversities and identification of particular proteins in teff, because SSPs are highly independent of environmental fluctuations [41]. The absence or presence and differences in band intensities can also be regarded as a basis for a possible polymorphism studies in teff since the type and amount of proteins in mature seeds are constant [42]. Further studies will be required for identification of seed type-specific proteins.

5. Conclusions

The yellowish color after alcohol extraction of teff flour can be an evidence for the presence of phenolic compounds and flavonoids, and some of the main compounds that give brown teff seed its color are alkaline soluble and are probably insoluble condensed tannins. Regarding nutritional qualities, Teff has superior essential amino acid content and balance compared to wheat, rice, barely and maize, with the brown seed having higher amino acid content than the brown. The addition of reducing agents to extraction solvents enhances prolamin yield during teff protein fractionation. Tert-butanol is more efficient than ethanol for extraction of teff prolamins. Glutelin is a major protein in teff seeds. The extracted glutelin content of white teff is significantly higher than that of brown teff. SDS-PAGE analysis can reveal genetic variability storage proteins among white and brown teff.

Author Contributions: Y.A.G. performed most of experiments and played a major role in preparing the manuscript; J.H.-I. ran instruments for sample preparation and helped data analysis in this study; K.Y.-S., K.M.-K. and K.K.-P. provided their expertise in designing the experiments, analyzing the data and preparing the manuscript.

Acknowledgments: In The first author gratefully acknowledges the support from National Institute for International Education Development (NIIED) for funding his study under the Korean Government Scholarship Program (KGSP) scheme.

Conflicts of Interest: The authors declare no conflict of interest.

References

1. Cannarozzi, G.; Plaza-Wüthrich, S.; Esfeld, K.; Larti, S.; Wilson, Y.; Girma, D.; Tadele, Z. Genome and transcriptome sequencing identifies breeding targets in the orphan crop tef (*Eragrostis tef*). *BMC Genomics* **2014**, *15*, 581. [CrossRef] [PubMed]

2. Mezemir, S. Probiotic potential and nutritional importance of teff (eragrostis tef (zucc) trotter) enjerra-a review. *Afr. J. Food Agric. Nutr. Dev.* **2015**, *15*, 9964–9981.

3. Shumoy, H.; Raes, K. Antioxidant Potentials and Phenolic Composition of Tef Varieties: An Indigenous Ethiopian Cereal. *Cereal Chem.* **2016**, *93*, 465–470. [CrossRef]

4. Adebowale, A.-R.A.; Emmambux, M.N.; Beukes, M.; Taylor, J.R.N. Fractionation and characterization of teff proteins. *J. Cereal Sci.* **2011**, *54*, 380–386. [CrossRef]

5. Baye, K. Teff: Nutrient Composition and Health Benefits. ESSP Working Paper 67. Available online: https://www.google.ch/url?sa=t&rct=j&q=&esrc=s&source=web&cd=1&ved=2ahUKEwjt7tfE09fiAhUiy4UKHZzTDtcQFjAAegQIAxAC&url=https%3A%2F%2Fwww.researchgate.net%2Fprofile%2FArvind_Singh56%2Fpost%2FWhat_does_the_global_production_and_consumption_of_teff_look_like%2Fattachment%2F59d64e4379197b80779a7a39%2FAS%253A492060228108288%2540149432770866%2Fdownload%2FBaye%2Bet%2Bal.%252C%2B2014.pdf&usg=AOvVaw0hh_EIUpZuCpZA_qq9qmDc (accessed on 05 January 2019).

6. Zhang, W.; Xu, J.; Bennetzen, J.L.; Messing, J. Teff, an Orphan Cereal in the Chloridoideae, Provides Insights into the Evolution of Storage Proteins in Grasses. *Genome Biol. Evol.* **2016**, *8*, 1712–1721. [CrossRef] [PubMed]

7. Gorinstein, S.; Pawelzik, E.; Delgado-Licon, E.; Haruenkit, R.; Weisz, M.; Trakhtenberg, S. Characterisation of pseudocereal and cereal proteins by protein and amino acid analyses. *J. Sci. Food Agric.* **2002**, *82*, 886–891. [CrossRef]

8. Mota, C.; Santos, M.; Mauro, R.; Samman, N.; Matos, A.S.; Torres, D.; Castanheira, I. Protein content and amino acids profile of pseudocereals. *Food Chem.* **2016**, *193*, 55–61. [CrossRef] [PubMed]

9. Tovar-Pérez, E.G.; Guerrero-Legarreta, I.; Farrés-González, A.; Soriano-Santos, J. Angiotensin I-converting enzyme-inhibitory peptide fractions from albumin 1 and globulin as obtained of amaranth grain. *Food Chem.* **2009**, *116*, 437–444. [CrossRef]

10. Leyva-Lopez, N.E.; Vasco, N.; Barba de la Rosa, A.P.; Paredes-Lopez, O. Amaranth seed proteins: Effect of defatting on extraction yield and on electrophoretic patterns. *Plant Foods Hum. Nutr.* **1995**, *47*, 49–53. [CrossRef] [PubMed]

11. Zheleznov, A.V.; Solonenko, L.P.; Zheleznova, N.B. Seed proteins of the wild and the cultivated Amaranthus species. *Euphytica* **1997**, *97*, 177–182. [CrossRef]

12. Tömösközi, S.; Gyenge, L.; Pelcéder, Á.; Abonyi, T.; Schönlechner, R.; Lásztity, R. Effects of Flour and Protein Preparations from Amaranth and Quinoa Seeds on the Rheological Properties of Wheat-Flour Dough and Bread Crumb. *Czech J. Food Sci.* **2011**, *29*, 109–116. [CrossRef]

13. El-Alfy, T.S.; Ezzat, S.M.; Sleem, A.A. Chemical and biological study of the seeds of Eragrostis tef (Zucc.) Trotter. *Nat. Prod. Res.* **2012**, *26*, 619–629. [CrossRef] [PubMed]

14. Zhu, F. Chemical composition and food uses of teff (*Eragrostis tef*). *Food Chem.* **2018**, *239*, 402–415. [CrossRef] [PubMed]

15. Tatham, A.S.; Fido, R.J.; Moore, C.M.; Kasarda, D.D.; Kuzmicky, D.D.; Keen, J.N.; Shewry, P.R. Characterisation of the Major Prolamins of Tef (*Eragrostis tef*) and Finger Millet (*Eleusine coracana*). *J. Cereal Sci.* **1996**, *24*, 65–71. [CrossRef]

16. Spackman, D.H.; Stein, W.H.; Moore, S. Automatic Recording Apparatus for Use in Chromatography of Amino Acids. *J. Anal. Chem.* **1958**, *30*, 1190–1206. [CrossRef]

17. Tella, A.F.; Ojehomon, O.O. An extraction method for evaluating the seed proteins of cowpea (Vigna unguiculata (L) Walp). *J. Sci. Food Agric.* **1980**, *31*, 1268–1274. [CrossRef]

18. Wallace, J.C.; Lopes, M.A.; Paiva, E.; Larkins, B.A. New Methods for Extraction and Quantitation of Zeins Reveal a High Content of gamma-Zein in Modified opaque-2 Maize. *Plant Physiol.* **1990**, *92*, 191–196. [CrossRef] [PubMed]

19. Van der Walt, W.H.; Schussler, L.; Van der Walt, W.H. Fractionation of proteins from low-tannin sorghum grain. *J. Agric. Food Chem.* **1984**, *32*, 149–154. [CrossRef]

20. Bradford, M.M. A Rapid and Sensitive Method for the Quantitation of Microgram Quantities of Protein Utilizing the Principle of Protein-Dye Binding. *J. Anal. Chem.* **1976**, *72*, 248–254. [CrossRef]

21. Laemmli, U.K. Cleavage of Structural Proteins during the Assembly of the Head of Bacteriophage T4. *Nature* **1970**, *227*, 680–685. [CrossRef]

22. Ravisankar, S., Abegaz, K., Awika, J.M. Structural profile of soluble and bound phenolic compounds in teff (*Eragrostis tef*) reveals abundance of distinctly different flavones in white and brown varieties. *Food Chem.* **2018**, *263*, 265–274. [CrossRef] [PubMed]

23. Antwi-Boasiako, C.; Animapauh, S.O. Tannin extraction from the barks of three tropical hardwoods for the production of adhesives. *J. Appl. Sci. Res.* **2005**, *8*, 2959–2965.

24. Chavan, J.K.; kadam, S.S.; Ghonsikar, C.P.; Salunkhe, D.K. Removal of tannins and improvement of in vitro protein digestibility of sorghum seeds by soaking in alkali. *J. Food Sci.* **1979**, *44*, 1319–1322. [CrossRef]

25. Tessari, P.; Lante, A.; Mosca, G. Essential amino acids: Master regulators of nutrition and environmental footprint? *Sci. Rep.* **2016**, *6*, 26074. [CrossRef] [PubMed]

26. Palombini, S.V.; Claus, T.; Maruyama, S.A.; Gohara, A.K.; Souza, A.H.P.; de Souza, N.E.; Visentainer, J.V.; Gomes, S.T.M.; Matsushita, M. Evaluation of nutritional compounds in new amaranth and quinoa cultivars. *Food Sci. Technol.* **2013**, *33*, 339–344. [CrossRef]

27. Mcdermott, E.E.; Pace, J. The content of amino-acids in white flour and bread. *Br. J. Nutr.* **1957**, *11*, 446–452. [CrossRef]

28. Mckevith, B. Nutritional aspects of cereals. *Food Nutr. Bull.* **2004**, *29*, 111–142. [CrossRef]

29. Thomas, R.; Bhat, R.; Kuang, Y.T. Composition of amino Acids, fatty acids, minerals and dietary fiber in some of the local and import rice varieties of Malaysia. *Int. Food Res. J.* **2015**, *22*, 1148–1155.

30. Ferreira, R.R.; Varisi, V.A.; Meinhardt, L.W.; Lea, P.J.; Azevedo, R.A. Are high-lysine cereal crops still a challenge? *Braz. J. Med. Biol. Res.* **2005**, *38*, 985–994. [CrossRef]

31. Dennison, C.; Lizette, M.; Che, S.P. *t*-Butanol: Nature's gift for protein isolation. *S. Afr. J. Sci.* **2000**, *96*, 159–160.

32. Gonzalez, J.A.; Konishi, Y.; Bruno, M.; Valoy, M.; Prado, F.E. Interrelationships among seed yield, total protein and amino acid composition of ten quinoa (*Chenopodium quinoa*) cultivars from two different agroecological regions. *J. Sci. Food Agric.* **2012**, *92*, 1222–1229. [CrossRef] [PubMed]

33. Mainieri, D.; Marrano, C.A.; Prinsi, B.; Maffi, D.; Tschofen, M.; Espen, L.; Vitale, A. Maize 16-kD γ-zein forms very unusual disulfide-bonded polymers in the endoplasmic reticulum: Implications for prolamin evolution. *J. Exp. Bot.* **2018**, *69*, 5013–5027. [CrossRef] [PubMed]

34. Nielsen, H.C.; Paulis, J.W.; James, C.; Wall, J.S. Extraction and structure studies on corn glutelin proteins. *Cereal Chem.* **1970**, *47*, 501.

35. Paulis, J.W.; Wall, J.S. Fractionation and properties of alkylated-reduced corn glutelin proteins. *BBA* **1971**, *251*, 57–69. [CrossRef]

36. Getz, E.B.; Xiao, M.; Chakrabarty, T.; Cooke, R.; Selvin, P.R. A Comparison between the Sulfhydryl Reductants Tris(2-carboxyethyl)phosphine and Dithiothreitol for Use in Protein Biochemistry. *Anal. Biochem.* **1999**, *273*, 73–80. [CrossRef] [PubMed]

37. Singh Khatkar, B. Structure and functionality of wheat gluten proteins. In Proceedings of the 2014 International Conference of Food Properties (ICFP 2014), Kuala Lumpur, Malaysia, 24–26 January 2014.

38. Coleman, J.M. Assessing the Potential Use of Teff as an Alternative Grain Crop in Virginia. MSc Thesis, Virginia Polytechnic Institute and State University, Blacksburg, VA, USA, April 2012.

39. Žilić, S.; Barać, M.; Pešić, M.; Dodig, D.; Ignjatović-Micić, D. Characterization of Proteins from Grain of Different Bread and Durum Wheat Genotypes. *Int. J. Mol. Sci.* **2011**, *12*, 5878–5894. [CrossRef]

40. Vollmann, J.; Fritz, C.N.; Wagentristl, H.; Ruckenbauer, P. Environmental and genetic variation of soybean seed protein content under Central European growing conditions. *J. Sci. Food Agric.* **2000**, *80*, 1300–1306. [CrossRef]

41. Javaid, A.; Ghafoor, A.; Anwar, R. Seed storage protein electrophoresis in groundnut for evaluating genetic diversity. *Pak. J. Bot.* **2004**, *36*, 25–29.

42. Sadia, M.; Malik, S.A.; Rabbani, M.A.; Pearce, S.R. Electrophoretic Characterization and the Relationship between some Brassica Species. *Electron. J. Bio.* **2009**, *5*, 1–4.

Commentary

The Effect of Processing on Digestion of Legume Proteins

Donata Drulyte and Vibeke Orlien *

Department of Food Science, Faculty of Science, University of Copenhagen, Rolighedsvej 26,
DK-1958 Frederiksberg C, Denmark; wnb429@alumni.ku.dk
* Correspondence: vor@food.ku.dk; Tel.: +45-35333226

Received: 15 May 2019; Accepted: 19 June 2019; Published: 24 June 2019

Abstract: The domestic processing methods, soaking, cooking (traditional, microwave, pressure), and baking and the industrial processing, autoclaving, baking, and extrusion are used to improve consumption of legumes. The growing awareness of both health and sustainability turns the focus on protein (bio)availability. This paper reports the effect of these processing methods on the legume protein digestibility. Overall, the protein digestibility increases after processing by the different methods. However, since both the type of legume and the applied methods differ it cannot be concluded which specific method is best for the individual legume type.

Keywords: legume protein; processing; digestibility

1. Introduction

In today's health-based consumer-focused world, more and more research is being conducted in order to obtain knowledge about the effects of animal- and plant-based diets on our health. It has generally been accepted that the consumption of meat can increase the incidence and prevalence of obesity, cardiovascular diseases and stroke, cancer, type 2 diabetes, and patient mortality. Contrarily, the plant-based diet has been accepted to lower the risk of these factors [1,2]. Consequently, an increasing percentage of people are changing their diets to become vegetarians, vegans, or flexitarians in order to lower their meat consumption and increase the amount of vegetables and fruits [3,4]. Yet, false information and health trends, which state that plant-based proteins are worse for your health, inadequate for building muscle, or are not proteinaceous enough, have become powerful and generally accepted misconceptions among some members of the population [5]. Research has managed to prove that legumes can be a good source of protein, though the bioavailability of animal proteins was still proven to be higher [6,7]. The recommended acceptable intake of proteins is around 0.8 g/kg for adults (defined as the average daily level of intake sufficient to meet the nutrient requirements of nearly all healthy people). Legumes have a great potential for delivering quality proteins, but raw legumes have a lower degree of bioavailability and, thereby, lower nutritional value as compared to other foodstuffs. Upon ending the last century, food was used to improve health, while entering the millennium a paradigm shift resulted in that our knowledge is now being used to improve foods in respect of healthiness. Both animal- and plant-based proteins have been widely investigated in order to understand the digestibility as such. Since it is well known that different processing procedures will affect both structure and functionality of proteins, it is anticipated that processing may also affect protein digestibility. This communication surveys the effect of different processing methods on legume protein digestion (PD) in order to contribute to the knowledge of protein digestibility. It is not a comprehensive review of all research studies on PD, but is based on the studies defined by investigations that contribute to provide an overview of the effect of processing on digestion of legume proteins. Thus, to "set the current scene" and give an outline of the challenges and future perspectives.

2. Processing Techniques

Consumption of raw legumes can be tedious and difficult, and, in worst case, inefficient with respect to amino acid absorption in the gut. Empirical, various domestic processing methods are used to ease the legume consumption per se, but without grasping the effect on protein digestibility as such. Several different types of legumes and of processing techniques have, thus, been investigated in order to evaluate the potential improvement of nutritional value and protein utilization. Processing is the action of performing a series of mechanical or thermal operations on food in order to change or preserve it. It may involve soaking, cooking, microwave irradiation, baking, pressure-cooking, autoclaving, and extrusion [8,9]. Since the processing methods are performed at different conditions, a brief description of the techniques that have been used is presented.

One of the commonly used pre-processing method is soaking. The most important parameters of soaking are: product:water ratio, temperature of the soaking water, and the duration. These parameters differ considerably depending on the type of legume, hence, the legumes are soaked in either hot or boiling tap or distilled water, with legume:water ratios varying from 1:1 to 1:5, and soaking durations ranging from 12 to 16 h. Traditional cooking is the simplest thermal processing method used. In the case of legumes, they are usually either placed in surplus warm water and slowly brought to boiling or directly put into boiling water (100 °C) in a pot. The time of the cooking process is a very important factor, which can differ depending on the legume type and if the legume seeds were soaked, dried, or not treated prior to cooking. Nevertheless, in a majority of the reviewed studies, the cooking time varied between 20 and 40 min. Microwave radiation is a more intense and, thereby, faster thermal method than conventional cooking. The thermal effects are related to the heat generated by the absorption of microwave energy by the water in the food matrix. Contrarily to cooking, microwave processing offers the opportunity to provide a more precise amount of energy to the food item. In addition, the heating effect of microwave can be tuned (energy dosage) and is considerably faster, almost immediate, resulting in reduction of cooking time. Similar to cooking, the microwave treatment time varies depending on type of legume and pre-treatment before microwave cooking. The capacity of microwave ovens differs due to different microwave frequency, thus delivering different energy doses, which will influence the treatment duration. In the studies herein, the doses of energy varies from 500 to 2000 J/g, usually being increased by 250 J/g. Pressure-cooking and autoclaving are two other high-intensity thermal methods. They are both based on exceeding normal, ambient boiling point of water in a sealed vessel (pressure cooker) or a pressure chamber (autoclave). The pressure cooker works by trapping the steam produced from boiling water of 121 °C inside the vessel. Many autoclaves work by subjecting the item to pressurized saturated steam at 121 °C. Thus, they both operate at 1.8 to 2.0 bar in order to obtain this temperature. The evaluated temperature and pressure generally increase reaction rates, thus the duration of pressure cooking and autoclaving is less than traditional cooking, though it varies between the two. Pressure cooking is usually performed in 5 to 15 min, whereas autoclave time may vary from seven to 60 min, both depending on the type and quantity of legume. The last two processing methods are based on milling the legumes into a flour prior to treatment. Baking is an old, traditional thermal method that uses dry air at ambient pressure to apply heat. Generally, a dough based on legume flour is made, rested for some time (leavening), and then baked at temperatures varying from about 180 to 220 °C for more than 30 min. Extrusion is also based on transforming legume flour into an edible food product. The extrudates are prepared using either a single- or twin-screw extruder that works under high temperature and high mechanical pressure. The main independent parameters, such as particle size of flour, feed rate, moisture content, barrel temperature, and screw speed, during the extrusion process, differ a lot depending on legume type and extruder parameters (type of mill, screen hole size, etc.). In the reviewed papers, moisture content was about 22–25%, while temperature and screw rotation speed varied considerably from 30 to 150 °C and from 100 to 650 rpm, respectively.

3. Protein Digestibility Methodologies

The major difference between PD assessment is if the measurement is conducted in vivo or in vitro. Furthermore, the digestion of proteins may be determined by a broad variety of methods within these two PD categories. In vivo methods are based on feeding trails using animals or humans. The most accurate result is obtained with controlled feeding experiments with animals, usually rodents, chicken, or pigs are used. However, care should be taken upon transferring results and conclusions from animal tests to human, since the human GI tract differs considerably from animals. On the one hand, human intervention studies are the golden method for assessing protein digestion and nutritional value. On the other hand, in vivo experiments are time consuming and costly, and difficult to control if the study is based on human intervention. Therefore, in vitro models simulating the human GI tract have been used. Contrarily to in vivo methods, in vitro methods are easier to control, rapid to conduct, and may be less expensive. The drawbacks among the in vitro methods are differences in the operation and the experimental parameters; as a result, there are considerable differences in the measured digestibility complicating comparison of results. In addition, the difficulties in accurately transferring and simulating the complex digestion process in the human GI tract complicates the performance of these laboratory experiments and caution must be taken upon interpretation of results. In the following, the various methods used to measure and assess legume protein digestibility in the surveyed papers are summarized.

3.1. In Vitro Protein Digestibility (IVPD)

The in vitro protein digestibility assay is the most used method for analyzing digestibility of protein samples. The analysis is carried out by first preparing a multi-enzyme solution, usually including trypsin, chymotrypsin, and peptidase, with some exceptions using single trypsin [10] or pepsin solution [11], or, sequentially, pepsin and pancreatin solutions [12–14]. Commonly, an aliquot of the enzyme solution is added to a solution of the sample at pH 8.0. The mixture of sample and enzyme solution is incubated for 10 min at 37 °C. The pH decrease during the incubation period is recorded, followed by calculations of the IVPD. The pH decrease is caused by the free amino acid carboxyl groups from the protein chain released by the proteolytic enzymes during the digestion.

3.2. Crude Protein (CP)

Crude protein is the quantity of proteins in a specific sample of feed or food. The crude protein is determined from the total content of nitrogen obtained by a total degradation of the proteins. The crude protein percentage is calculated using the Kjeldahl conversion factor of 6.25 [15,16].

3.3. Ileal Digestability (AID, SID)

The ileal digestibility of the proteins under investigation is based on intervention study. Thus, different protein diets (usually with similar crude protein and trypsin inhibitor activity levels) are fed to animals, usually rats, pigs, or chickens, living under controlled conditions. The feeding trials may run from 10 to 20 days depending on study and animal. Afterwards, the animals are euthanized and the ileal digesta is collected by mildly flushing the digesta with clean distilled water and the amount of crude protein or amino acids is measured. Overall, the ileal protein digestibility is the difference between the amount of protein or amino acid ingested by the animal and the amount of protein or amino acid in the ileal digesta outflow. However, the calculation of the ileal digestibility depends on which amount of the ileal protein/amino acid outflow is used in the actual calculation and can be calculated as apparent (AID), true (TID), or standardized (SID) ileal digestibility. The AID for a certain amino acid is computed by subtracting the total ileal outflow of that specific amino acid from the quantity consumed by the animal. The values for SID are computed similarly to the values for AID, apart from the fact that the basal ileal endogenous amino acid (AA) losses (IAAend) are subtracted from the ileal outflow [17–19].

4. Effect of Processing on Protein Digestibility

Legumes are rich sources of proteins (18–41%) and are important raw food materials worldwide. The proteins provide the essential amino acids necessary for maintaining body muscle and growth. However, legumes, and, thereby their proteins, are not similar. Moreover, the nutritional quality of proteins is not solely dictated by the AA composition, but important factors like digestion rate and digestibility in the GI tract, which in turn are determined by the protein structure and enzyme accessibility, are of outmost importance. The protein digestibility is affected by both endogenous and exogenous factors. Endogenous factors relate to the protein as such, that is, protein structural characteristics and how and to what extent food processing may affect this structure. Exogenous factors are related to the food matrix, and include protein interactions with other compounds like carbohydrates, lipids, and especially anti-nutritional factors (ANF). ANFs include the anti-nutritional proteins like trypsin inhibitors and lectins and the anti-nutritional chemicals like tannins, phytates, and polyphenols. Therefore, it may beforehand be expected that both different types of legume protein and different processing may result in a very diverse digestibility. Indeed, Table S1 (Supplementary Materials) shows the difference in protein digestibility of legumes using various processing methods.

4.1. Cooking

The simplest processing technique, conventional cooking, is the most studied method with 12 studies as seen in Table S1. Four different types of legumes, beans, peas, lentils, and chickpeas, but of various cultivars, were investigated. Most of the investigations were reporting in vitro protein digestibility, while only few studies reported protein content changes [15] and in vivo digestion [16,20]. The protein content in the flour of unprocessed bean, pea, and lentil is rather similar, ranging from 24.8% to 28.7% [15]. However, comparing legume subtypes that underwent cooking, the effect on the protein content differed considerably (Table S1). Unprocessed bean (*Phaseolus vulgaris*) with protein contents of 24.8% (Raba) and 26.2% (Warta) significantly decreased to 23.0% and 21.3%, respectively. The protein content in cooked and uncooked peas (*Pisum Savitum*) yielded quite different results, since the content in the Milwa cultivar increased, while the Medal's protein content decreased [15]. The lentil cultivars (*Lens culinaris*) demonstrated both the highest and a significant increase in protein content after cooking. However, digestibility was not investigated [15]. For all cooked legumes, IVPD increased significantly, as seen in Table S1. For the flour of eight different species of unprocessed peas (*Pisum sativum* L.) the IVPD ranged from 79.9% to 83.5%, while the IVPD of the processed samples varied between 85.9% and 86.8% [21]. The IVPD of 83.61% for unprocessed chickpea (*Cicer arietinum* L.) increased to 88.52% after cooking for 90 min [22]. The time of cooking was found to be important for the IVPD, since Habiba et al. [23] found an increasing IVPD upon increasing cooking time. Interestingly, the increased IVPD was concomitant with a decrease in total protein content. This decrease in total crude proteins was suggested to be a result of leaching of water-soluble proteins during cooking [23]. A similar explanation is likely for the reported decreases in protein content for some of the legumes mentioned above [15]. The cooking resulted in improved IVPD of lentils, chickpea, peas, and soybean, but soaking the legumes prior to heating did not result in consistent significant effects [24]. Similarly, cooking of three different varieties of kidney beans significantly increased the IVPDs, while pre-soaking did not have any major effect [25]. Moreover, soaking in alkaline solution (sodium bicarbonate, pH 8.2) did not improve protein digestibility. However, Embaby [26] found that soaking cooked bitter and sweet lupin seeds for 96 and 24 h, respectively, further improved the IVPD.

In same line, the in vivo digestibility differed according to the lentil type, temperature, and time applied. Digestibility (SID) of unprocessed full-fat soybeans (FFSB) was 46% [16]. Incremental increase of the cooking temperature and duration caused a correlated increase in the soybean's SID. Hence, beans cooked at 80 °C for one min had a SID of 52%, whereas FFSB processed at 100 °C for six or 16 min had a SID of 73% and 80%, respectively, Table S1. Similarly, the cooking of peas (*Pisum sativum* L.) prior using as a diet resulted in an increase of true digestibility (TD) (79.8%) as compared to raw pea diet with TD of 74.7% [20].

For centuries, prior human consumption legume seeds have been soaked and thermally treated by conventional cooking due to the simplicity in the execution and equipment. However, the drawbacks of cooking are a fairly uncontrolled and non-adjustable process and the potential loss of valuable nutrients like vitamins. Therefore, other processing techniques are investigated in order to optimize the protein digestion by better control of the heating process.

4.2. Microwave Cooking

IVPD of three faba bean cultivars before processing were 46.0%, 52.2%, 51.5% for Windsor White, Bacchus, and Basta, respectively, and, thereby, characterized by a markedly lower IVPD compared to other varieties of seeds reported in former section [27]. Generally, treatment with microwave radiation resulted in an increase in protein digestibility of all bean types (Table S1). The lowest amount of energy (500 J/g) caused a significant increase in protein digestibility from 46.0%, 52.2%, and 51.5% to 57.1%, 68.0%, and 53.2%, respectively [27]. Further increase in energy to 1000 J/g significantly improved protein digestibility to 76.5% for Windsor White, 76.1% for Bacchus, and 78.2% for Basta. However, more energy input (1250, 1500, 1750 J/g) during microwave cooking did not significantly affect the protein digestibility further [27]. The authors concluded that microwave processing at 1000 J/g is optimal for the protein digestibility of faba beans [27]. Soaking is a traditional domestic method for preparing seeds for further processing. Embaby [26] investigated the reverse situation, thus soaking of bitter and sweet lupin seeds for 96 and 24 h, respectively, after microwave treatment. It was found that microwave processing significantly improved the IVPD by 2.5% and 1.5% compared to the raw seeds (from 78.55% to 80.40% for bitter lupin and from 79.46% to 80.67% for sweet lupin, Table S1). Thus, soaking following microwave cooking further improved the IVPD for bitter lupin seeds, while no significant increase was found for the sweet lupin seeds [26]. It is noted that the considerably longer soaking time (96 h) for the bitter lupin seeds compared to 24 h for the sweet lupin seeds may have a major impact on this observation, though the author did not comment on this.

4.3. Pressure Cooking

Pressure cooking is another common domestic treatment method utilizing the high-energy input to shorten processing time. Protein digestibility of four unprocessed moth bean cultivars differed significantly between 70.3–74.6% for the local variety to new varieties [12]. After pressure-cooking, IVPD increased to around 78% for the local, Jwala, and RMO 225, while RMO 257, with the highest raw IVPD, also had the highest cooked IVPD of 82.4% (Table S1). Soaking prior pressure cooking in addition to reducing processing time, positively affected bean protein digestibility, since the IVPDs were improved by 14–16% [12]. Pressure cooking of peas was also found to improve the digestibility of proteins compared to digestibility of raw peas [23]. The IVPDs resulting from standard pressure cooking were at the same level as the IVPDs obtained after cooking irrespective of processing times, though a shorter treatment time was necessary for pressure cooking, while slightly higher than the IVPDs after microwave treatment (Table S1) [23].

4.4. Autoclaving

Autoclaving of beans and peas significantly reduced the content of crude proteins compared to the raw legumes, while the content in the lentils was unaffected (Table S1) [15]. Hence, the autoclaving treatment of raw legumes had different effect on the protein content compared to the cooking process. Small differences between the two heat treatments were also observed for the Milwa pea and the two lentil cultivars [15]. The two studies reporting the effect of autoclaving of yellow peas [28] and soybean [15] on in vivo protein digestibility showed an increasing tendency, Table S1. Though the legume types, the processing conditions, and the animal model differed, it seems that the rather harsh heat treatment of the plant material prior inclusion in the diet improved the protein digestion in the GI tract. Similarly, autoclaving generally increased the IVPD significantly for all types of legumes investigated, expect for faba bean and lentil [29] (Table S1). Hence, unambiguous improvement of protein digestibility was not obtained after autoclaving compared to cooking. The more harsh and intense heat treatment by autoclaving did seemingly not always have a positive effect. Furthermore, the increase of autoclaving time from 10 to 90 min significantly reduced the IVPD of four different legumes [11].

4.5. Baking

House and co-workers have investigated the effect of baking of pea and lentil flours on protein digestibility [30,31]. Apparently, the process of mixing, kneading, rising, and baking of the dough reduced the IVPD compared to cooking, except for the red lentil (Table S1). Thus, the authors concluded that, for home preparation of these legumes, cooking is more advantageous than baking [31].

4.6. Extrusion

Extrusion is a thermal process with high energy efficiency due to high shear and compression, and probably the most severe thermal treatment method. Nevertheless, the extrusion process had a positive effect on the nutritional value of legumes. Extruding the flour of common beans, pea seed, faba, and kidney beans significantly increased the IVPD up to 87% [14,32,33]. The in vitro digestibility results were supported by in vivo feeding experiments. Hence, feeding chicken with extruded peas or kidney beans improved the effect on the apparent ileal digestibility of crude protein [34,35]. The AIDs of CP for unprocessed and extruded pea seeds (*Pisum sativum L., Tarachalska cv.*) were 74.3% and 85.9%, respectively, thus extrusion increased protein digestibility [21]. Inclusion of extruded kidney bean (100–300 g/kg) in broiler diets increased AID to 85.5–85.9% compared to AID of 77.23–79.03% with feed based on raw kidney beans [35].

4.7. New Processing Methods

High pressure and ultrasound are non-thermal technologies known to only affect non-covalent interactions in macromolecules. Therefore, the possible pressure-induced protein unfolding may enhance the access of the digestive enzymes to the cleavage sites in the protein, thus improving digestibility. However, as seen in Table S1, the effects on IVPD of lentils, chickpeas, peas, and soybean after high pressure or ultrasound treatment were inconsistent or insignificant compared to the protein digestibility of the raw legumes [10,24].

4.8. Factors Affecting PD

Various processing techniques affect legume PD to various extents compared to the PD of raw legumes. Generally, the IVPD increases after processing by different heating methods. The improvement in PD was attributed to protein denaturation. Thus, the increased digestibility was ascribed to the resulting heat-induced denaturation of the proteins, thereby enhancing accessibility of susceptible sites to proteolysis. On the other hand, digestibility can be compromised by protein aggregation due to the thermal treatment [29]. A consequence of protein denaturation is increased opportunity for various intra- and intermolecular interactions, especially disulfide bridges between amino acids containing free thiol groups. Crosslinked, aggregated proteins are less accessible to digestive enzymes because of a different localization of amino acid residues specific for protease action resulting in lack of PD improvement. It is noted that the degree of amino acid reactivity and extent of aggregation differs under various conditions and not all amino acids participate in protein crosslinking irrespective of the processing condition. In addition, the native protein conformation of the primary legume proteins, globulins and albumins, may also affect the PD as such. The albumins have a compact globular structure stabilized by a large number of disulfide bonds, thus possessing an inherent structural hindrance limiting enzyme access [21]. In that respect, it was observed that an increased proportion of globulins and decreased proportion of albumins in pea seeds could contribute to an increased IVPD [21]. This may indicate that the ratio of albumin:globulin also has an influence on the IVPD of legumes. It is generally accepted that the abundance of ANF in plant protein sources contributes to the lower digestibility compared to typical mammal proteins. Phytic acid, tannins, and polyphenols may interact with protein to form complexes by cross-linking with the proteins, resulting in a decreased protein solubility and making these protein complexes less susceptible to proteolytic attack in the GI tract. Partly elimination of tannins and phytic acid was observed after different thermal treatments (Table S1). The reduction of reactive tannins and phytic acid would result in less protein complexing and creating more space within the matrix, which increased the accessibility of the enzyme resulting in improved IVPD [11]. Trypsin inhibitors may interfere with the action of proteolytic enzymes in the GI tract by forming inactive complexes of trypsin and chymotrypsin. However, since trypsin inhibitors are heat-labile compounds, thermal processing should be an effective method for reducing its activity. Indeed, a marked decrease in trypsin inhibitors after different thermal processing was reported in many of the studies (Table S1) and a complete inactivation was also found for peas and kidney beans [23,25]. However, Embaby found that some treatments resulted in an increased level of anti-nutrients, but still obtained improved IVPD [26]. Thus, he suggested that ANFs are not solely responsible for lowering IVPD, and factors like cell wall rigidity and fiber content may influence the protein digestibility as such.

In conclusion, the improvement of PD of legume proteins for all reported process methods is explained by the structural disintegration of the native protein concomitant with the reduction or even removal of anti-nutrients.

5. Challenges and Future Perspective

The nutritional value of plant-based food is dependent on the amount of proteins, the specific distribution of the amino acids, and, most importantly, the bioavailability of these amino acids. Reduced protein quality will compromise the nutritional value, therefore maintaining quality and stability upon processing of plant-base materials is of utmost importance. From nature, proteins are not equally digestible, as their proteolytic susceptibility varies due to different three-dimensional structures according to their origin. This inherent difference in digestibility across legume sources is mirrored into the effect of processing methods. The benefits of food processing covers preservation, palatability, functional properties, or added convenience, but the impact of processing are often disconnected to the nutritional status of the proteins. Surely, the ultimate goals are plant-based products with high protein quality to be enjoyed as part of the consumer's healthy diet and lifestyle. Assessing the quality of protein in relation to nutritional value means to determine the capacity of the proteins to satisfy the metabolic demand for amino acids. If perfect, a measure of the nutritional quality of dietary protein should provide the real protein digestibility and, thereby, enable the prediction of the overall efficiency of protein utilization. At present, various in vitro and in vivo protein digestibility methods exist, and neither of them can be used to accurately measure protein digestibility in absolute terms. An important conceptual difficulty is that proteins differ in their digestibility and the human GI tracts differ in their ability to digest proteins. Therefore, there is still much to learn and improve regarding methodologies for measuring and monitoring the protein digestibility before the "true" effect of processing on the important amino acids can be established. The golden goal for optimizing nutrition may be monitoring the individual protein digestive capacity in the individual body. However, even if food producers used the perfect PD method to optimize PD of a food product, the product must still appeal to consumers for success on the market. It is recognized that an increasing number of food companies are producing successful plant-based products providing variety, taste, and nutritional value to consumers.

Supplementary Materials: The following are available online at http://www.mdpi.com/2304-8158/8/6/224/s1, Table S1. Protein digestibility (PD) of legumes using various processing methods.

Author Contributions: Conceptualization, V.O.; investigation, D.D.; writing—original draft preparation, D.D.; writing—review and editing, V.O.

Funding: This research received no external funding.

Conflicts of Interest: The authors declare no conflicts of interest.

References

1. Riboli, E.; Hunt, K.J.; Slimani, N.; Ferrari, P.; Norat, T.; Fahey, M.; Saracci, R. European prospective investigation into cancer and nutrition (EPIC): Study populations and data collection. *Public Health Nutr.* **2002**, *5*, 1113–1124. [CrossRef] [PubMed]
2. Carlos, S.; De La Fuente-Arrillaga, C.; Bes-Rastrollo, M.; Razquin, C.; Rico-Campà, A.; Martínez-González, M.; Ruiz-Canela, M. Mediterranean diet and health outcomes in the SUN cohort. *Nutrients* **2018**, *10*, 429. [CrossRef] [PubMed]
3. Radnitz, C.; Beezhold, B.; DiMatteo, J. Investigation of lifestyle choices of individuals following a vegan diet for health and ethical reasons. *Appetite* **2015**, *90*, 31–36. [CrossRef] [PubMed]
4. Rosenfeld, D.L.; Burrow, A.L. Vegetarian on purpose: Understanding the motivations of plant-based dieters. *Appetite* **2017**, *116*, 456–463. [CrossRef] [PubMed]
5. DellaBartolomea, M. 4 Myths about Plant-Based Protein. 2018. Available online: https://www.nutrasciencelabs.com/blog/4-myths-about-plant-based-protein (accessed on 21 June 2019).
6. Hoffman, J.R.; Falvo, M.J. Protein—Which is best? *J. Sports Sci. Med.* **2004**, *3*, 118–130. [PubMed]
7. Tomé, D. Digestibility issues of vegetable versus animal proteins: Protein and amino acid requirements—Functional aspects. *Food Nutr. Bull.* **2013**, *34*, 272–274. [CrossRef] [PubMed]
8. Öste, R.E. Digestibility of processed food protein. *Adv. Exp. Med. Biol.* **1991**, *289*, 371–388. [PubMed]
9. Johnson, P.E. Effect of food processing and preparation on mineral utilization. *Adv. Exp. Med. Biol.* **1991**, *289*, 483–498.
10. Linsberger-Martin, G.; Weiglhofer, K.; Thi Phuong, T.; Berghofer, E. High hydrostatic pressure influences antinutritional factors and in vitro protein digestibility of split peas and whole white beans. *LWT Food Sci. Technol.* **2013**, *51*, 331–336. [CrossRef]
11. Rehman, Z.; Shah, W. Thermal heat processing effects on antinutrients, protein and starch digestibility of food legumes. *Food Chem.* **2005**, *91*, 327–331. [CrossRef]
12. Negi, A.; Boora, P.; Khetarpaul, N. Starch and protein digestibility of newly released moth bean cultivars: Effect of soaking, dehulling, germination and pressure cooking. *Nahrung* **2001**, *45*, 251–254. [CrossRef]
13. Khatoon, N.; Prakash, J. Nutritional quality of microwave-cooked and pressure-cooked legumes. *Int. J. Food Sci. Nutr.* **2004**, *55*, 441–448. [CrossRef] [PubMed]
14. Batista, K.A.; Prudêncio, S.H.; Fernandes, K.F. Changes in the functional properties and antinutritional factors of extruded hard-to-cook common beans (Phaseolus vulgaris, L.). *J. Food Sci.* **2010**, *75*, 286–290. [CrossRef] [PubMed]
15. Piecyk, M.; Wołosiak, R.; Drużynska, B.; Worobiej, E. Chemical composition and starch digestibility in flours from polish processed legume seeds. *Food Chem.* **2012**, *135*, 1057–1064. [CrossRef] [PubMed]
16. Kaewtapee, C.; Eklund, M.; Wiltafsky, M.; Piepho, H.P.; Mosenthin, R.; Rosenfelder, P. Influence of wet heating and autoclaving on chemical composition and standardized ileal crude protein and amino acid digestibility in full-fat soybeans for pigs. *J. Anim. Sci.* **2017**, *95*, 779–788. [PubMed]
17. Jansman, A.J.M.; Smink, W.; van Leeuwen, P.; Rademacher, R. Evaluation through literature data of the amount and amino acid composition of basal endogenous crude protein at the terminalileum of pigs. *Anim. Feed Sci. Technol.* **2002**, *98*, 49–60. [CrossRef]
18. GFE. Standardised precaecal digestibility of amino acids in feedstuffs for pigs—Methods and concepts. In *Mitteilungen des Ausschusses für Bedarfsnormen der Gesellschaft für Ernährungsphysiologie (Communications of the Committee for Requirement and Standards of the Society of Nutrition Physiology*; Martens, H., Ed.; DLG-Verlag: Frankfurt a. M., Germany, 2005; pp. 185–205.
19. Stein, H.H.; Fuller, M.F.; Moughan, P.J.; Sève, B.; Mosenthin, R.; Jansman, A.J.M.; Fernández, J.A.; de Lange, C.F.M. Definition of apparent, true, and standardized ileal digestibility of amino acids in pigs. *Livest. Sci.* **2007**, *109*, 282–285. [CrossRef]
20. Nagra, S.A.; Bhatty, N. In vivo (rat assay) assessment of nutritional improvement of peas (*Pisum sativum* L.). *East Mediterr. Health J.* **2007**, *13*, 646–653. [PubMed]

21. Park, S.J.; Kim, T.W.; Baik, B.K. Relationship between proportion and composition of albumins, and in vitro protein digestibility of raw and cooked pea seeds (*Pisum sativum* L.). *J. Sci. Food Agric.* **2010**, *90*, 1719–1725. [CrossRef]

22. el-Adawy, T.A. Nutritional composition and antinutritional factors of chickpeas (*Cicer arietinum* L.) undergoing different cooking methods and germination. *Plant Foods Hum. Nutr.* **2002**, *57*, 83–97. [CrossRef]

23. Habiba, R. Changes in anti-nutrients, protein solubility, digestibility, and HCl-extractability of ash and phosphorus in vegetable peas as affected by cooking methods. *Food Chem.* **2002**, *77*, 187–192. [CrossRef]

24. Han, I.; Swanson, B.; Baik, B. Protein digestibility of selected legumes treated with ultrasound and high hydrostatic pressure during soaking. *Cereal Chem.* **2007**, *84*, 518–521. [CrossRef]

25. Shimelis, E.; Rakshit, S. Effect of processing on antinutrients and in vitro protein digestibility of kidney bean (*Phaseolus vulgaris* L.) varieties grown in East Africa. *Food Chem.* **2007**, *103*, 161–172. [CrossRef]

26. Embaby, H. Effect of soaking, dehulling, and cooking methods on certain antinutrients and in vitro protein digestibility of bitter and sweet lupin seeds. *Food Sci. Biotechnol.* **2010**, *19*, 1055–1062. [CrossRef]

27. Pysz, M.; Polaszczyk, S.; Leszczyńska, T.; Piątkowska, E. Effect of microwave field on trypsin inhibitors activity and protein quality of broad bean seeds (Vicia faba var. major). *Acta Sci. Pol. Technol. Aliment.* **2012**, *11*, 193–198. [PubMed]

28. Frikha, M.; Valencia, D.G.; de Coca-Sinova, A.; Lázaro, R.; Mateos, G.G. Ileal digestibility of amino acids of unheated and autoclaved pea protein concentrate in broilers. *Poult. Sci.* **2013**, *92*, 1848–1857. [CrossRef] [PubMed]

29. Carbonaro, M.; Cappelloni, M.; Nicoli, S.; Lucarini, M.; Carnovale, E. Solubility–Digestibility relationship of legume proteins. *J. Agric. Food Chem.* **1997**, *45*, 3387–3394. [CrossRef]

30. Nosworthy, M.G.; Franczyk, A.J.; Medina, G.; Neufeld, J.; Appah, P.; Utioh, A.; Frohlich, P.; House, J.D. Effect of processing on the in Vitro and in Vivo protein quality of yellow and green split peas (*Pisum sativum*). *J. Agric. Food Chem.* **2017**, *65*, 7790–7796. [CrossRef] [PubMed]

31. Nosworthy, M.G.; Medina, G.; Franczyk, A.J.; Neufeld, J.; Appah, P.; Utioh, A.; Frohlich, P.; House, J.D. Effect of processing on the in vitro and in vivo protein quality of red and green lentils (Lens culinaris). *Food Chem.* **2018**, *240*, 588–593. [CrossRef]

32. Alonso, R.; Aguirre, A.; Marzo, F. Effects of extrusion and traditional processing methods on antinutrients and in vitro digestibility of protein and starch in faba and kidney beans. *Food Chem.* **2000**, *68*, 159–165. [CrossRef]

33. Alonso, R.; Grant, G.; Dewey, P.; Marzo, F. Nutritional assessment in vitro and in vivo of raw and extruded peas (*Pisum sativum* L.). *J. Agric. Food Chem.* **2000**, *48*, 2286–2290. [CrossRef] [PubMed]

34. Hejdysz, M.; Kaczmarek, S.A.; Rutkowski, A. Effect of extrusion on the nutritional value of peas for broiler chickens. *Arch. Anim. Nutr.* **2016**, *70*, 364–377. [CrossRef] [PubMed]

35. Arija, I.; Centeno, C.; Viveros, A.; Brenes, A.; Marzo, F.; Illera, J.C.; Silvan, G. Nutritional evaluation of raw and extruded kidney bean (Phaseolus vulgaris L. var. pinto) in chicken diets. *Poult. Sci.* **2006**, *85*, 635–644. [CrossRef] [PubMed]

Table 1. Protein digestibility (PD) of legumes using various processing methods. ↓ indicates decreased PD. ↑ indicates increased PD. In addition, the effect on anti-nutritional factors (ANF) is given when provided in the references. Empty cell denotes that information was not provided.

Process	Type of Legume (Cultivar)	Specification	Soaked	Unsoaked	Process Conditions (T; t; ΔH)	Methodology [1]	PD Unprocessed (%)	PD Processed (%)	Effect	Reference
	Bean flour (Phaseolus vulgaris)	Raba	*		30 min	CP	24.8	23.0	↓ (significantly)	[15]
		Warta	*		40 min		26.2	21.3		
	Pea (Pisum sativum)	Milwa	*		40 min		25.9	27.6	↑ (significantly)	
		Medal	*		40 min		28.4	27.5	↓ (significantly)	
	Lentil (Lens culinaris)	Anita	*		20 min		28.7	29.2		
		Tina	*		20 min		28.6	30.0	↑ (significantly)	
	Pea (Pisum sativum L.) flour	Yellow-green (Windham)	*		98 °C, 30 min	IVPD (trypsin, chymotrypsin, and peptidase, 10 min)	79.9	85.9	↑ (significantly)	[21]
		Yellow-green (Spector)	*		98 °C, 30 min		81.3	85.9		
		Green (Stirling)	*		98 °C, 30 min		82.0	85.9		
		Yellow (Fallon)	*		98 °C, 30 min		82.6	86.5		
		Pale green (Supra)	*		98 °C, 30 min		82.4	86.0		
		Yellow (Shawnee)	*		98 °C, 30 min		83.5	86.8		
		Green (Lifter)	*		98 °C, 30 min		83.5	86.3		
		Yellow-green (DSP)	*		98 °C, 30 min		82.3	86.8		
		Mean	*		98 °C, 30 min		82.2	86.3		
	Full-fat soybean (FFSB)			*	80 °C, 1 min	SID (pig) of CP	46	52	↑ (Reduced trypsin inhibitors)	[16]
				*	100 °C, 6 min			73		
				*	100 °C, 16 min			80		
	Pea (Pisum sativum) flour	Green pea	*		brought to a boil and maintained until done, 25–35 min	IVPD (trypsin, chymotrypsin and protease, 10 min)		86.56		[30]
		Yellow pea	*					86.75		
	Lentil (Lens culinaris) flour	Red lentil	*		brought to a boil (100 °C) and maintained, 25–35 min	IVPD (trypsin, chymotrypsin and protease, 10 min)		84.67		[31]
		Green lentil	*					84.03		
	Pea (Pisum sativum L.)			*	100 °C, 40 min—high heat; 30 min—simmered	TD	74.7	79.8	↑ (significantly)	[20]
	Chickpea (Cicer arietinum L.)		*		100 °C, 90 min	IVPD (trypsin, pancreatin, 10 min)	83.61	88.52	↑ (significantly) (Reduced trypsin inhibitors, tannins, phytic acid)	[22]
	Bitter lupin seed (Lupinus termis)		*	*	Cooked in distilled water (100 °C) 1:10 (w/v), 40 min	IVPD (multienzyme assay, 10 min)	78.55	80.73	↑ (significantly) (Reduced trypsin inhibitors, phytic acid (except sweet lupin), tannins)	[26]
			(after cooking)					89.72		
	Sweet lupin seed (Lupinus albus)			*			79.46	84.35		
			* (after cooking)					86.09		
	Vegetable pea (Pisum sativum L.)			*	Cooked in water (1:2 seed:water) at 100 °C — 20 min	IVPD (trypsin, α-chymotrypsin, peptidase, 10 min)	73.5	76.0	↑ (Reduced trypsin inhibitors, phytic acid, tannins)	[23]
				*	30 min			77.2		
				*	40 min			78.3		

Table 1. *Cont.*

Process	Type of Legume (Cultivar)	Specification	Soaked	Unsoaked	Process Conditions (T; t; ΔH)	Methodology [1]	PD Unprocessed (%)	PD Processed (%)	Effect	Reference
Cooking	Lentil (*Lens culinaris*)	*Pardina*		*	98 °C for 30 min	IVPD (trypsin, chymotrypsin, and peptidase)	79.1	83.6	↑ (significantly)	[24]
			*, 3 h					83.3	↑	
			*, 12 h					83.6		
		Crimson		*			79.4	82.2	↑	
			*, 3 h					83.1	(significantly)	
			*, 12 h					83.7		
	Chickpea (*Cicer arietinum* L.)			*			74.3	82.5	↑ (significantly)	
			*, 3 h					82.9	↑ (significantly)	
			*, 12 h					82.4	↑	
	Pea (*Pisum sativum*)	*Yellow*		*			82.0	84.0	↑ (significantly)	
			*, 3 h					84.5	↑ (significantly)	
			*, 12 h					84.3	↑	
		Green		*			82.6	84.1	↑ (significantly)	
			*, 3 h					84.5	↑	
			*, 12 h					83.6		
	Soybean (*Glycine max*)			*			71.8	83.8	↑ (significantly)	
			*, 3 h					85.1		
			*, 12 h					85.0		
	Black gram	*Punjab91*	*		Boiled, seed:water ratio of 1:5	IVPD (pepsin-HCl, 24 h)	34.8	65.0	↑ (significantly) (Reduced phytic acid, tannins)	[11]
	Chickpea	*CP-98*	*				36.0	69.0		
	Lentil	*Nayyab2002*	*				37.6	72.7		
	Kidney bean	*Red (Chkwal99)*	*				33.8	64.1		
		White (WK-70)	*				34.0	63.6		
	Kidney bean (*P. vulgaris* L.)	*Roba*		*	1:3 (*w/v*), 97 °C, 35 min	IVPD (multienzyme assay)	80.66	87.11	↑ (significantly) (Reduced trypsin inhibitors, phytic acid, tannins)	[25]
			*					90.31		
			NaHCO₃					90.34		
		Awash		*			71.14	78.28		
			*					78.98		
			NaHCO₃					78.88		
		Beshbesh		*			65.63	71.55		
			*					73.52		
			NaHCO₃					72.85		

Table 1. *Cont.*

Process	Type of Legume (Cultivar)	Specification	Soaked	Unsoaked	Process Conditions (T; t; ΔH)	Methodology [1]	PD Unprocessed (%)	PD Processed (%)	Effect	Reference
Microwave cooking	Bean (*vicia faba var. major*)	*Windsor White*	*		500 J/g	IVPD (peptidase, trypsin, chymotrypsin)	46.0	57.1	↑ (significantly) (Reduced trypsin inhibitors)	[27]
			*		750 J/g			69.4		
			*		1000 J/g			76.5		
			*		1250 J/g			78.7		
			*		1500 J/g			79.0		
			*		1750 J/g			78.8		
			*		2000 J/g			78.8		
		Bacchus	*		500 J/g		52.2	68.0		
			*		750 J/g			73.2		
			*		1000 J/g			76.1		
			*		1250 J/g			78.4		
			*		1500 J/g			78.9		
			*		1750 J/g			79.0		
			*		2000 J/g			78.8		
		Basta	*		500 J/g		51.5	53.2		
			*		750 J/g			64.1		
			*		1000 J/g			78.2		
			*		1250 J/g			78.8		
			*		1500 J/g			79.8		
			*		1750 J/g			81.2		
			*		2000 J/g			81.0		
	Bean	Broad bean (*Vicia faba*)	*			IVPD (pepsin, pancreatin)		77.4		[13]
		Field bean (*Dolichos lablab*)	*					71.2		
		Green gram (*Phaseolus aureus Roxb*)	*					78.1		
		Horse gram (*Dolichos biflorus*)	*					69.8		
		French bean (*Phaseolus vulgaris*)	*					74.6		
	Lentil	*Lens esculenta*	*					78.7		
	Chickpea	*Bengal gram (Cicer arietinum)*	*					74.1		
	Pea	*Cowpea (Vigna catjang)*	*					73.7		
	Chickpea (*Cicer arietinum* L.)		*		on high level, 15 min	IVPD (trypsin, pancreatin, 10 min)	83.61	89.40	↑ (Reduced trypsin inhibitors, tannins, phytic acid)	[22]

Table 1. *Cont.*

Process	Type of Legume (Cultivar)	Specification	Soaked	Unsoaked	Process Conditions (T; t; ΔH)	Methodology [1]	PD Unprocessed (%)	PD Processed (%)	Effect	Reference
	Bitter lupin seed (*Lupinus termis*)			*				80.54		
	Bitter lupin seed (*Lupinus termis*)		* (76 h after MW)		seed:water ratio 1:10, for 6 min	IVPD (multienzyme method, 10 min)	78.55	85.75	↑ (significantly) (Reduced trypsin inhibitors, phytic acid, tannins)	[26]
	Sweet lupin seed (*Lupinus albus*)			*				80.67	↑ (significantly) (Reduced trypsin inhibitors, tannins)	[26]
	Sweet lupin seed (*Lupinus albus*)		* (24 h after MW)				79.46	81.51	↑ (Reduced trypsin inhibitors, tannins)	[26]
	Vegetable pea (*Pisum sativum* L.)			*	2450 MHz — 4 min	IVPD (trypsin, α-chymotrypsin, peptidase, 10 min)	73.5	74.2	↑ (Reduced trypsin inhibitors, tannins, and phytic acid (only at 12 min))	[23]
	Vegetable pea (*Pisum sativum* L.)			*	2450 MHz — 8 min			75.1		
	Vegetable pea (*Pisum sativum* L.)			*	2450 MHz — 12 min			75.5		
Pressure cooking	Moth bean (*Phaseolus aconitifolius* Jacq.)	*Local*		*	15 min	IVPD (pepsin, pancreatic)	70.3	78.1		[12]
Pressure cooking	Moth bean (*Phaseolus aconitifolius* Jacq.)	*Local*	*		10 min		70.3	81.4		[12]
Pressure cooking	Moth bean (*Phaseolus aconitifolius* Jacq.)	*Jwala*		*	15 min		71.9	78.9		[12]
Pressure cooking	Moth bean (*Phaseolus aconitifolius* Jacq.)	*Jwala*	*		10 min		71.9	81.9		[12]
Pressure cooking	Moth bean (*Phaseolus aconitifolius* Jacq.)	*RMO 225*		*	15 min		72.3	78.2		[12]
Pressure cooking	Moth bean (*Phaseolus aconitifolius* Jacq.)	*RMO 225*	*		10 min		72.3	83.9		[12]
Pressure cooking	Moth bean (*Phaseolus aconitifolius* Jacq.)	*RMO 257*		*	15 min		74.7	82.4		[12]
Pressure cooking	Moth bean (*Phaseolus aconitifolius* Jacq.)	*RMO 257*	*		10 min		74.7	85.2		[12]
Pressure cooking	Bean	Broad bean (*Vicia faba*)	*			IVPD (pepsin, pancreatin)		80.0		[13]
Pressure cooking	Bean	Field bean (*Dolichos lablab*)	*					75.9		[13]
Pressure cooking	Bean	Green gram (*Phaseolus aureus* Roxb)	*					81.9		[13]
Pressure cooking	Bean	Horse gram (*Dolichos biflorus*)	*					78.4		[13]
Pressure cooking	Bean	French bean (*Phaseolus vulgaris*)	*					78.9		[13]
Pressure cooking	Lentil	*Lens esculenta*	*					82.6		[13]
Pressure cooking	Chickpea	Bengal gram (*Cicer arietinum*)	*					80.1		[13]
Pressure cooking	Pea	Cowpea (*Vigna catjang*)	*					80.5		[13]
Pressure cooking	Vegetable pea (*Pisum sativum* L.)			*	120 °C — 10 min	IVPD (trypsin, α-chymotrypsin, peptidase, 10 min)	73.5	77.4	↑ (Reduced trypsin inhibitors, tannins, and phytic acid (only at 20 min))	[23]
Pressure cooking	Vegetable pea (*Pisum sativum* L.)			*	120 °C — 15 min			78.3		[23]
Pressure cooking	Vegetable pea (*Pisum sativum* L.)			*	120 °C — 20 min			78.12		[23]

Table 1. *Cont.*

Process	Type of Legume (Cultivar)	Specification	Soaked	Unsoaked	Process Conditions (T; t; ΔH)		Methodology [1]	PD Unprocessed (%)	PD Processed (%)	Effect	Reference
Autoclaving	Bean flour (*Phaseolus vulgaris*)	Raba		*	121 °C, 7 min		CP	24.8	21.8	↓ (significantly)	[15]
		Warta		*	121 °C, 12 min			26.2	21.0		
	Pea flour (*Pisum sativum*)	Milwa		*	121 °C, 12 min			25.9	24.4		
		Medal		*	121 °C, 12 min			28.4	27.1		
	Lentil flour (*Lens culinaris*)	Anita		*	121 °C, 7 min			28.7	28.2	↓	
		Tina		*	121 °C, 7 min			28.6	28.7	↑	
	Yellow pea flour (*Pisum sativum*, L.)	Concentrate 1		*	108 °C, 8 min		AID and SID (chicken) of CP	AID 80.1	88.1	↑ (Reduced trypsin inhibitors)	[28]
								SID 88.7	95.0	↑ (significantly) (Reduced trypsin inhibitors)	
		Concentrate 2		*	108 °C, 8 min			AID 84.3	87.2	(Reduced trypsin inhibitors)	
								SID 91.6	95.0		
	Full-fat soybean (FFSB)			*	110 °C	15 min	SID (pig) of CP	46	82	↑ (Reduced trypsin inhibitors)	[16]
				*		30 min			83		
				*		45 min			84		
				*		60 min			82		
	Chickpea (*Cicer arietinum* L.)		*		121 °C, 35 min		IVPD (trypsin, pancreatin, 10 min)	83.61	89.96	↑ (Reduced trypsin inhibitors, tannins, phytic acid)	[22]
	Faba bean (*Vicia faba* L.)		*		120 °C, 20 min		IVPD (trypsin, chymotrypsin and peptidase, 10 min, protease, 9 min)	83.07	79.57	↓ (significantly)	[29]
	Lentil (*Lens culinaris Medikus*)		*					82.50	81.66	↓	
	Chickpea (*Cicer arietinum* L.)		*					78.28	83.16	↑ (significantly)	
	Dry white bean (*Phaseolus vulgaris* L.)		*					73.76	80.53	↑ (significantly)	
	Bitter lupin seed (*Lupinus termis*)		* (after treatment)	*	121 °C, 20 min		IVPD (multienzyme assay, 10 min)	78.55	82.84	↑ (significantly) (Reduced trypsin inhibitors, phytic acid (except sweet lupin), tannins)	[26]
									91.53		
	Sweet lupin seed (*Lupinus albus*)		* (after treatment)	*				79.46	83.93		
									85.95		

Table 1. *Cont.*

Process	Type of Legume (Cultivar)	Specification	Soaked	Unsoaked	Process Conditions (T; t; ΔH)	Methodology [1]	PD Unprocessed (%)	PD Processed (%)	Effect	Reference
	Black gram	Punjab91	*		121 °C, 10 min	IVPD (pepsin-HCl, 24 h)	34.8	68.0	↑ (significantly) (Reduced phytic acid, tannins)	[11]
					121 °C, 20 min			64.5		
					121 °C, 40 min			62.8		
					121 °C, 60 min			62.0		
					121 °C, 90 min			61.8		
					128 °C, 20 min			62.4		
	Chickpea	CP-98	*		121 °C, 10 min		36.0	72.5		
					121 °C, 20 min			68.5		
					121 °C, 40 min			67.0		
					121 °C, 60 min			65.7		
					121 °C, 90 min			64.9		
					128 °C, 20 min			65.0		
	Lentil	Nayyab2002	*		121 °C, 10 min		37.6	76.0		
					121 °C, 20 min			72.0		
					121 °C, 40 min			70.8		
					121 °C, 60 min			68.0		
					121 °C, 90 min			67.8		
					128 °C, 20 min			68.0		
	Kidney bean	Red (Chkwal99)	*		121 °C, 10 min		33.8	68.3		
					121 °C, 20 min			64.0		
					121 °C, 40 min			62.4		
					121 °C, 60 min			61.0		
					121 °C, 90 min			60.5		
					128 °C, 20 min			61.8		
		White (WK-70)	*		121 °C, 10 min		34.0	69.8		
					121 °C, 20 min			63.0		
					121 °C, 40 min			61.4		
					121 °C, 60 min			60.0		
					121 °C, 90 min			59.9		
					128 °C, 20 min			61.0		
	Kidney bean (P. vulgaris L.)	Roba		*	1:3 (w/v), 121 °C, 30 min	IVPD (multienzyme assay)	80.66	86.31	↑ (significantly) (Reduced trypsin inhibitors, phytic acid, tannins)	[25]
			*					92.76		
			NaHCO₃					92.84		
		Awash		*			71.14	76.12		
			*					82.53		
			NaHCO₃					81.82		
		Beshbesh		*			65.63	69.58		
			*					74.77		
			NaHCO₃					74.83		
Baking	Dry pea flour (Pisum sativum)	Yellow pea		*	198.3 °C, 198.3 °C, and 165.6 °C, 35 min	IVPD (trypsin, chymotrypsin, and protease, 10 min)		82.49		[30]
		Green pea		*	29 min			82.22		
	Lentil flour (Lens culinaris)	Red lentil		*	198.3 °C, 198.3 °C, and 165.6 °C, 35 min	IVPD (trypsin, chymotrypsin and protease, 10 min)		85.03		[31]
		Green lentil		*				79.33		

Table 1. *Cont.*

Process	Type of Legume (Cultivar)	Specification	Soaked	Unsoaked	Process Conditions (T; t; ΔH)	Methodology [1]	PD Unprocessed (%)	PD Processed (%)	Effect	Reference
Extrusion	Dry pea flour (*Pisum sativum*)	*Yellow pea*		*	30–50 °C, 70–90 °C, and 100–120 °C	IVPD (trypsin, chymotrypsin and protease, 10 min)		86.93		[30]
		Green pea		*				90.0		
	Lentil flour (Lens culinaris)	*Red lentil*		*	30–50 °C, 70–90 °C, and 100–120 °C	IVPD (trypsin, chymotrypsin and protease, 10 min)		88.01		[31]
		Green lentil		*				84.30		
	Pea (*Pisum sativum* L., Tarachalska cv.)			*	135 ± 10 °C	AID of CP	74.3	85.9	↑ (significantly) (Reduced trypsin inhibitors, phytic acid)	[33]
	Common Bean flour (*Phaseolus vulgaris*, L.)	*BRS pontal (carioca)*		*	150 °C	IVPD (pepsin, pancreatin)	28.16	48.52	↑ (significantly) (Reduced trypsin inhibitors, phytic acid)	[14]
		BRS grafite (black)		*			28.62	52.80		
	Pea flour (*Pisum sativum* L.)			*	145 °C	IVPD (trypsin,α-chymotrypsin, peptidase)	839 g/kg	874 g/kg	↑ (significantly) (Reduced trypsin inhibitors, phytic acid, tannins, polypehnols)	[35]
	Kidney Bean flour (*Phaseolus vulgaris* L. var. Pinto)	*100 g/kg of extruded kidney bean (EKB)*		*	150 °C	AID of CP	89.60 (without kidney bean), 79.03 (100 g/kg raw KB), 79.53 (200 g/kg raw KB), 77.23 (300 g/kg raw KB)	85.50	↑ (significantly)	[34]
		200 g/kg of EKB		*				85.50		
		300 g/kg of EKB		*				85.90		
	Bean flour	Common bean (*Phaseolus vulgaris*, L.)		*	152 °C, 156 °C	IVPD (trypsin,α-chymotrypsin, peptidase, 10 min)	68.1	83.0	↑ (significantly) (Reduced trypsin inhibitors, phytic acid, tannins, polypehnols)	[32]
		Broad bean (*Vicia faba*)		*			70.8	87.4		

Table 1. *Cont.*

Process	Type of Legume (Cultivar)	Specification	Soaked	Unsoaked	Process Conditions (T; t; ΔH)	Methodology [1]	PD Unprocessed (%)	PD Processed (%)	Effect	Reference
New processing methods	Lentil (*Lens culinaris*)	*Pardina*	*1, 1.5 h		47 kHz	IVPD (trypsin, chymotrypsin, and peptidase)	79.1	79.8	↑ (significantly)	[24]
			*1, 3 h					80.1		
			*2, 0.5 h		621 MPa			79.1	↑ (significantly)	
			*2, 1 h					79.7		
		Crimson	*1, 1.5 h		47 kHz		79.4	80.2	↑ (significantly)	
			*1, 3 h					80.9		
			*2, 0.5 h		621 MPa			79.8	↑ (significantly)	
			*2, 1 h					81.0		
	Chickpea (*Cicer arietinum* L.)		*1, 1.5 h		47 kHz		74.3	74.1		
			*1, 3 h					74.5		
			*2, 0.5 h		621 MPa			74.9	↑ (significantly)	
			*2, 1 h					76.2		
	Pea (*Pisum sativum*)	*Yellow*	*1, 1.5 h		47 kHz		82.0	81.2	↓ (significantly)	
			*1, 3 h					81.2		
			*2, 0.5 h		621 MPa			81.2		
			*2, 1 h					81.3		
		Green	*1, 1.5 h		47 kHz		82.6	81.9	↓ (significantly)	
			*1, 3 h					82.0		
			*2, 0.5 h		621 MPa			81.2		
			*2, 1 h					82.0		
	Soybean (*Glycine max*)		*1, 1.5 h		47 kHz		71.8	72.1	↑	
			*1, 3 h					71.6	↓	
			*2, 0.5 h		621 MPa			72.2	↑ (significantly)	
			*2, 1 h					73.1		

Table 1. *Cont.*

Process	Type of Legume (Cultivar)	Specification	Soaked	Unsoaked	Process Conditions (T; t; ΔH)	Methodology [1]	PD Unprocessed (%)	PD Processed (%)	Effect	Reference
	Dry split pea (*Pisum sativum*)				350 MPa, 45 min, 40 °C	IVPD (trypsin)	82.3	80.7 (mean of all three)	↓ (Reduced trypsin inhibitors, phytic acid)	[10]
					600 MPa, 60 min, 60 °C			85.8	↑ (Reduced trypsin inhibitors, phytic acid)	
					100 MPa, 60 min, 20 °C			78.6	↓ (Reduced trypsin inhibitors, phytic acid)	
					600 MPa, 30 min, 20 °C			82.3	(Reduced trypsin inhibitors, phytic acid)	
					100 MPa, 60 min, 60 °C			79.2	↓ (Reduced trypsin inhibitors, phytic acid)	
					600 MPa, 30 min, 60 °C			83.3	↑ (Reduced trypsin inhibitors, phytic acid)	
					100 MPa, 30 min, 60 °C			79.6		
					600 MPa, 60 min, 20 °C			82.1	↓ (Reduced trypsin inhibitors, phytic acid)	
					100 MPa, 30 min, 20 °C			79.9		
	White beans (*Phaseolus vulgaris*)				350 MPa, 45 min, 40 °C		69.1	68.0 (mean of all three)	↓ (Reduced trypsin inhibitors, phytic acid)	
					600 MPa, 60 min, 60 °C			75.1	↑ (Reduced trypsin inhibitors, phytic acid)	
					100 MPa, 60 min, 20 °C			68.5		
					600 MPa, 30 min, 20 °C			69.0		
					100 MPa, 60 min, 60 °C			67.5		
					600 MPa, 30 min, 60 °C			68.8	↓ (Reduced trypsin inhibitors, phytic acid)	
					100 MPa, 30 min, 60 °C			68.8		
					600 MPa, 60 min, 20 °C			68.9		
					100 MPa, 30 min, 20 °C			63.8		

[1] CP: Srude protein, IVPD: In vitro protein digestibility, SID: standardized ileal digestibility, TD: True digestibility, AID: Apparent ileal digestibility. *1 Soaked under ultrasound, *2 Soaked under high pressure.

MDPI

St. Alban-Anlage 66

4052 Basel

Switzerland

Tel. +41 61 683 77 34

Fax +41 61 302 89 18

www.mdpi.com

Foods Editorial Office

E-mail: foods@mdpi.com

www.mdpi.com/journal/foods